MW00856035

GEOLOGY
UNDERFOOT
in Southern Utah

Richard L. Orndorff,
Robert W. Wieder,
and David G. Futey

2006
MOUNTAIN PRESS PUBLISHING COMPANY
Missoula, Montana

© 2006 by Richard L. Orndorff, Robert W. Wieder, and David G. Futey
All rights reserved

On front cover: Moon over Park Avenue,
Arches National Park (from photo by David G. Futey)
Pictured on back: Paria Canyon hoodoo (David G. Futey)

The Geology Underfoot series presents geology with a hands-on,
get-out-of-the-car approach. A formal background in geology
is not required for enjoyment.

Geology Underfoot is a registered trademark of
Mountain Press Publishing Company.

All photographs by the authors unless otherwise credited.

Library of Congress Cataloging-in-Publication Data
Orndorff, Richard L.
  Geology underfoot in southern Utah / Richard L. Orndorff, Robert W.
Wieder, and David G. Futey.
    p. cm.
  Includes index.
  ISBN-13: 978-0-87842-517-4 (pbk. : alk. paper)
1. Geology—Utah. 2. Formations (Geology)—Utah. 3. Geoparks—Utah.
4. National parks and reserves—Utah. 5. Parks—Utah. I. Wieder, Robert
W., 1958- II. Futey, David G., 1959- III. Title.
  QE169.O76 2006
  557.92—dc22
                                                          2005032973

PRINTED IN THE UNITED STATES OF AMERICA
BY FC PRINTING, SALT LAKE CITY, UTAH

Mountain Press Publishing Company
P.O. Box 2399 • Missoula, MT 59806
(406) 728-1900 • www.mountain-press.com

Men come and go, cities rise and fall, whole civilizations appear and disappear—the earth remains, slightly modified. The earth remains, and the heartbreaking beauty where there are no hearts to break. Turning Plato and Hegel on their heads I sometimes choose to think, no doubt perversely, that man is a dream, thought an illusion, and only rock is real. Rock and sun.

Edward Abbey, *Desert Solitaire*

Nothing lives long. Only the earth and the mountains . . .

White Antelope (Cheyenne), 1864

Delicate Arch

CONTENTS

PREFACE
In Flagrante Delicto

A while back, one of this book's authors, Bob Wieder, took a long hot ride down a rough dirt road in southern Utah to visit a place called Hole in the Rock. In January 1880, Brigham Young sent a group of Mormon settlers out from St. George, Utah, to found a settlement in the southeastern part of the state. The journey went smoothly until they were confronted with the 2,000-foot-deep Colorado River canyon. Being industrious and no strangers to hard work, these pioneers chiseled and blasted their way through rock down to the river. After six weeks of unbelievable labor, they lowered wagons using ropes and oxen and finally managed to ferry their possessions across the muddy Colorado.

Like the settlers, Bob ran into an obstacle, a piece of road too tough even for his beat-up old pickup. He parked and walked the final two miles to the slice in the canyon wall called Hole in the Rock and clambered down the long, steep slope to the river (now submerged by Lake Powell). He found a small cove, ate lunch, and went for a swim. By the time he climbed back to the rim, the sun was high and hot. Rather than hiking back to the truck in the midday sun, he relaxed in a shaded rock overhang to let the heat pass. The view was amazing. Nothing but burnt red sandstone and blue sky, with the Straight Cliffs to the west and snowcapped Navajo Mountain to the south. Bob hadn't seen another person all day so, being in a clothing-optional situation, he stripped down to his skin. There he sat, naked, sipping water, eating peanuts, and watching the shadows change with the passing sun. It was so quiet he could hear the sound of swallows' wings as they flew by his sheltered alcove. He drifted off to sleep.

Suddenly the earth shook with a strange, unsettling sound. Disturbed, Bob crawled out from his hideaway. Just as he stood and stretched, a large helicopter filled with tourists flew over a sandstone knob and dipped down into the canyon directly toward him. Bob saw numerous wide eyes and mouths and, seconds too late, a mother struggling to cover her daughter's eyes. It must have been a sight—a scraggly, unshaven, sunburned, and very naked geologist smiling and waving in the late afternoon sun. The helicopter made a quick turn to the west and was gone, leaving Bob alone with the frantic swallows, swirling dust, and chopping echoes.

The heli-tourists certainly wanted to experience southern Utah, but what did they really hear and smell and see? The overwhelming throb of engines, the odor of exhaust, and a panorama of colors and shapes that changed so quickly that nothing could be remembered with any clarity. That's very different from the sound of birds' wings cutting through air, the tang of desert sagebrush, and the rough embrace of warm sandstone on your back. You don't have to strip naked (and we'd prefer Bob follow this advice in the future, at least when we're around), but by all means, slow down, get out of the car, and take a walk. Yep, listen to the wind, smell the dirt, and stub a toe.

THE GEOLOGY OF SOUTHERN UTAH

Spend some time with a map of southern Utah (defined in this book as that part of Utah mostly south of I-70, a convenient dividing line) and you'll discover place-names that reflect the diverse geology and history of the state. Box Death Hollow, Devils Garden, Burnt Gulch, and Bloody Hands Gap attest to difficult times and hard ends. Some place-names are imbued with mystery—The Maze, Dark Canyon, and No Man's Mesa. In this dry land, apparently not all water is sweet; consider the Dirty Devil River, Poison Springs Canyon, and Bitter Creek. Kodachrome Basin, Vermilion Cliffs, Coral Pink Sand Dunes, and Rainbow Mountain promise landscapes saturated with color, while other place-names accentuate the bawdy (Cad's Crotch, Nipple Spring, Nipple Creek, and Mollie's Nipple). Native Americans lived here first, and their ongoing influence is recorded in names such as Mount Kinesava, Kaiparowits Plateau, and Hoskinnini Mesa. Mormon settlers also added their signatures to Lees Ferry, Pioneer Gap, and Mount Brigham. The names on the map promise intrigue and grandeur, and the landscape delivers. Southern Utah's color and diverse landforms are a function of the underlying geology, and we'll share the details of some of that geology with you in this book.

The geology of a landscape is the end result of a series of events—both gradual and cataclysmic—that have occurred since the formation of our planet some 4.6 billion years ago. There are many ways to visualize just how long a period of time this is. If you're a Green Bay Packers fan like coauthor Dave Futey, who actually attends winter games at Lambeau Field, you might imagine compressing the entire history of the earth into a 100-yard football field. You catch the ball at your own goal line and take off upfield. At your 30-yard line you pass the oldest rocks we know, our first clues to our planet's past. Living organisms first arose in the sea at the 42-yard line, but you wouldn't see land plants and animals until reaching your opponent's 9-yard line, a total distance of 91 yards.

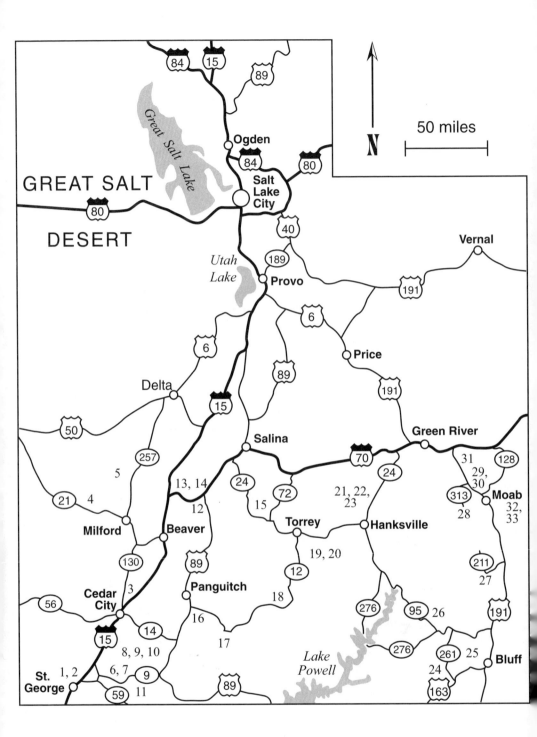

You would sprint past dinosaurs, which dominated the earth between the 8- and 1-yard lines. The Rocky Mountains formed at the 1-yard line, our human ancestors first walked upright at the 4-inch line, and the most recent continental ice sheets receded from the upper Midwest at 0.07 inch. The great civilizations of Egypt, Greece, and Rome rose and fell between 0.003 inch and the goal line. You push forward and fall into the end zone, exhilarated but exhausted by your sprint through geologic time.

More formally, geologists divide geologic time into eons, eras, periods, and epochs, with each interval defined by significant events in the earth's history. The Phanerozoic Eon, for example, begins about 540 million years ago and encompasses only the most recent 12 percent of geologic time. *Phanerozoic* means "apparent life," and the eon's beginning is marked by the appearance of multicelled organisms with hard parts (such as shells and teeth) that are preserved as fossils. The Phanerozoic Eon is subdivided into the *Paleozoic, Mesozoic,* and *Cenozoic* Eras, Latin for "ancient life," "middle life," and "new life," respectively. The Paleozoic Era witnessed a succession of dominant organisms, from invertebrates to fish to amphibians. Reptiles first evolved in the Paleozoic but didn't dominate the world until the Mesozoic, when dinosaurs flourished. The boundary between the Mesozoic and Cenozoic is defined by the mass extinction of

EON	ERA	PERIOD	EPOCH	
Phanerozoic	Cenozoic	Quaternary	Holocene Pleistocene	PRESENT 1.8 MILLION YEARS
		Tertiary	Pliocene Miocene Oligocene Eocene Paleocene	65 MILLION YEARS
	Mesozoic	Cretaceous Jurassic Triassic		245 MILLION YEARS
	Paleozoic	Permian Pennsylvanian Mississippian Devonian Silurian Ordovician Cambrian		540 MILLION YEARS
Precambrian				4.5 BILLION YEARS

Geologic time scale

many organisms, including the dinosaurs. The Cenozoic, known as the age of mammals, again saw a succession of dominant organisms.

The fossils found in rock allow us to establish a sequence of ages. With fossiliferous rocks, we can determine that one rock is older than another based on the fossils found therein, but fossils alone provide only a relative age. To assign a numeric value to the age of a rock, we must look elsewhere. Absolute ages for rocks can be estimated using radiometric dating, a method that relies on the presence and steady decay of radioactive elements. We discuss this method in detail in vignette 16, where the ages of displaced lava flows are used to estimate the rate of offset for the Sevier fault.

In addition to having created a classification system for geologic time, geologists have also created a classification system for landscapes. This system divides North America into physiographic provinces, which are regions that share prominent physical characteristics. Different regions in Utah are characterized by particular landforms due to consistent underlying geologic processes.

Southwestern Utah lies in the Basin and Range province, a region of north-south trending mountain ranges separated by deep valleys. To understand why this region exhibits its distinctive landforms, we first have to understand the basics of plate tectonics. Most of us know that the earth is made up of the core, the mantle, and the crust. The core contains mostly iron and nickel (very heavy elements) and provides the earth with its magnetic field. The mantle is dense and thick, and the crust is a very thin layer of relatively light rock. Plate tectonics, a theory widely accepted since the early 1970s, introduces two additional layers. Together, the uppermost mantle and crust behave as a brittle solid, which is called the lithosphere. The portion of upper mantle lying immediately beneath the lithosphere, called the asthenosphere, is ductile and capable of flow. Movement within the asthenosphere drives movement of the lithosphere, which is broken up into numerous plates. Plates interact with each other at their boundaries, moving together, moving apart, or sliding by one another, in the process giving rise to most of the volcanoes and earthquakes on our planet.

The Basin and Range province lies within the North American Plate and reflects a developing divergent plate tectonic boundary. Semi-molten rock in the asthenosphere is welling up beneath the Basin and Range, causing broad doming and thinning of the lithosphere. The upwelling material then moves to the east and west, literally pulling the continent apart. (There is no immediate danger to inhabitants, however, as plate movement averages only centimeters per year.) This stretching of the crust causes pieces of the lithosphere to fracture and fall downward to form the desert valleys that alternate with mountains in the Basin and Range. If this process continues, a great rift valley will eventually form, deepening over time until it fills with water to create a new ocean.

Physiographic provinces and relief of Utah

Southeastern Utah lies in the Colorado Plateau, an uplifted region that, unlike the Basin and Range province, is relatively undeformed. Some geologic terms are poetic, but others are just cumbersome. The word that describes this sort of uplift, *epeirogeny*, obviously doesn't fall within the poetic category. The Colorado Plateau, really a group of plateaus separated from one another by north-trending faults or broad folds, is bordered by centers of volcanism, where molten rock has migrated to the surface through paths of weakness created by faulting. The transition zone between the Basin and Range province and the Colorado Plateau exhibits characteristics of both provinces.

Southern Utah also sits within two hydrologic provinces, defined by the behavior of flowing water. Southeastern Utah is part of the immense Colorado River watershed, which extends through seven states and two countries. Prior to its disturbance by humans, the Colorado River freely flowed into the Sea of Cortez in the Gulf of California. This river has justly been called the lifeblood of the West; without it, cities like Phoenix, Los Angeles, and Las Vegas would be dusty crossroads instead of the crowded urban areas they are today.

Southwestern Utah, on the other hand, lies within the easternmost part of the Great Basin, which encompasses the central southwestern United States. Most river systems eventually empty into the sea, but the Great Basin is a closed basin, meaning that none of its rivers exit to the sea. It's bounded by the Sierra Nevada to the west and the Rocky Mountains to the east, and highlands to the north and south prevent the escape of water in those directions as well. This creates an immense bowl from which water can only escape by evaporation into the atmosphere. The Sevier River, for example, flows north in central Utah, then makes an abrupt turn west and south to spill into Sevier Lake, a salty depression with no outlet that only contains water during a small fraction of the year.

In the chapters that follow, we'll discuss sites that reflect many of the geologic processes that have impacted southern Utah. We'll discuss canyons, buttes, mesas, arches, and natural bridges, and we'll contemplate trackways of the mighty dinosaurs who ruled the world during Mesozoic time. We'll consider the remains of gigantic eruptions that left behind only scorched earth and smoking ash, weigh evidence for ancient meteorite impacts, and explore dry lakebeds. We'll guide you to recent fault scarps and huge landslides, as well as glacial features; you may find the latter surprising, given the warm, dry climate that currently prevails in this region. We'll also talk about silver- and uranium-bearing rock and associated mining activities in Utah. As you visit the sites described in this book, we want to share some ideas about the link between ancient and modern processes and the landscapes they've created. We hope you'll come away with an appreciation for the region's geology while maintaining a sense of wonder at the colorful majesty of southern Utah.

It's important to be well prepared when visiting some of the sites in this book. Summer temperatures in southern Utah often soar over 100 degrees Fahrenheit, and the mountains can be quite cool and wet throughout the year. Remember that you're traveling in a desert and always carry more water than you think you'll need. Never enter enclosed canyons if it looks like rain, and pay attention to weather reports from upstream areas that contribute water to canyons you plan to explore. All of the sites described here can be accessed by passenger car, but do inquire locally about conditions on unpaved roads. Intense thunderstorms can wash away sections of road. Most sites in national and state parks will require entrance fees—if you plan to visit several parks, it's probably more cost-effective to buy a season pass. Wear sturdy shoes and bring your camera and lots of film—southern Utah is one of the most photogenic landscapes on earth. Most of all, have fun!

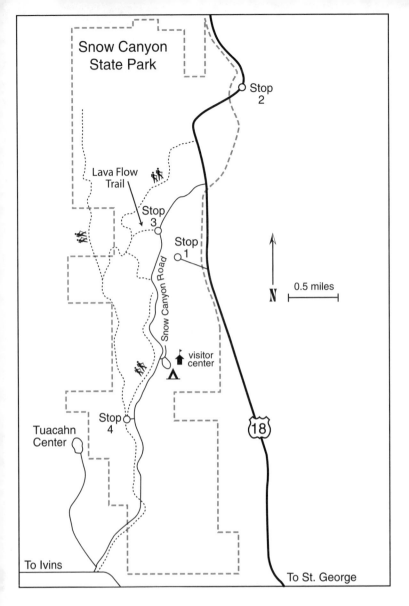

GETTING THERE

Snow Canyon is easily accessible from I-15. To get to stop 1, take exit 6 (Utah 18) and proceed north through St. George about 3 miles to the intersection with Snow Canyon Parkway. A sign here instructs you to turn left to Snow Canyon, but don't make the turn. Instead, continue north on Utah 18 for approximately 7 more miles to a sign for the Snow Canyon Overlook. Turn left onto the dirt road and follow it to the end (about 0.25 mile). To get to stop 2, the cinder cone parking area, return to Utah 18, turn left, and proceed north about 2.25 miles. The parking area is on the right just south of Diamond Valley Drive. To get to stop 3, carefully turn around, go back south on Utah 18 about 1.5 miles, and turn right onto Snow Canyon Road (Utah 300). Pass through the north park entrance and continue another 0.7 mile south to a parking area on the right for the Lava Flow Trail. The trail is 2 miles round-trip. To reach stop 4, continue south on Snow Canyon Road another 2.1 miles and turn right into a small gravel pull-out. Hackberry Wash is to the south (left).

1 READING A LAVA LANDSCAPE
Snow Canyon State Park

Snow Canyon State Park, tucked away in the southwest corner of Utah, is a small geological gem of a park. Here, deep canyons have been cut down into the red and white Navajo sandstone (see vignette 8), creating a dramatic backdrop for hiking, camping, and other recreational activities. Three times in the recent past ("recent" in geologic time, meaning the last couple million years), these canyons have played host to lava flows, and each of these flows has in turn caused Snow Canyon to migrate west. This series of events has resulted in a classic example of a phenomenon called inverted topography, in which an older rock unit sits above a younger one.

Our investigation into the unique geology of Snow Canyon begins at stop 1, a vantage point for all three basalt flows. To the east, back across the highway, the Lava Ridge flow caps a cliff of white Navajo sandstone. At about 1.4 million years in age, this is the oldest of the three flows. It erupted from a group of vents and weathered cinder cones along Lava Ridge, about 2.5 miles northeast of this overlook. The black rock on which you're standing at stop 1 is basalt from the Snow Canyon Overlook flow. This flow erupted about 1.1 million years ago and probably originated from vents near the Pine Valley Mountains, about 5 to 10 miles northeast of the park. In the canyon below, you can see the Santa Clara flow. Sometime between 20,000 and 10,000 years ago, this flow erupted from fissures at the base of the cinder cones at the head of Snow Canyon, north of where you're standing. The lava then flowed southward, spilling into Snow Canyon and covering the canyon floor. The large, rounded mounds of Navajo sandstone protruding above the flow, called turtlebacks, are remnants of high ridges the lava flowed around.

1

Small cinder cone volcano at the north end of Snow Canyon State Park

Think about the sequence we just described. The oldest of the three flows, the Lava Ridge flow, is at the highest elevation, with the subsequent flows forming a steplike pattern down to the west with the youngest, the Santa Clara flow, at the bottom. This sequence seems counterintuitive. After all, one of the most basic of geologic laws, the principle of superposition, tells us that in any sequence of undisturbed rocks, the oldest layer is at the bottom, with the sequence getting progressively younger as you go up. One generally thinks in terms of sedimentary rocks when applying this law, but it also holds true for volcanic rocks. Let's try to piece together the series of events that led to the inverted sequence at Snow Canyon.

When the Lava Ridge flow erupted about 1.4 million years ago, it spread out over the Navajo sandstone and hardened to a thick black sheet of basalt that buried much of the preexisting topography. With any former drainages now filled with basalt, a type of rock very resistant to erosion, water had to establish new drainage patterns. The softer Navajo sandstone at the western edge of the flow offered much less resistance to the downcutting action of water, so the drainage shifted west. Over the next 300,000 years, this new drainage developed into an ancestral version of Snow Canyon.

About 1.1 million years ago, a second episode of volcanic activity sent another thick river of lava down the ancestral Snow Canyon and filled it to the level at which you now stand. Because the earlier Lava Ridge flow had precipitated the carving of a deep new canyon along its west flank, this younger Snow Canyon Overlook flow, which filled the new canyon, sits stratigraphically lower than the older flow.

With its drainage once again filled with basalt, surface water had to cut a new path into the softer sandstone along the new flow's western edge. This process, which continued for over a million years, formed modern Snow Canyon, which lies below you to the west. At one time, Snow Canyon was probably V-shaped and significantly deeper than it is today. However, sometime between 20,000 and 10,000 years ago, renewed volcanic activity sent the Santa Clara flow down the canyon, filling it up to the level you see and creating a mostly flat canyon floor. Only a few high points, the sandstone turtlebacks, protrude above the flow. Here we have another case of inverted topography: the older Snow Canyon Overlook flow is situated above the younger Santa Clara flow in the canyon below. If you look at the western edge of the Santa Clara

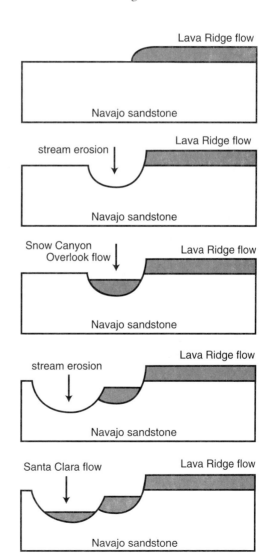

Alternating lava flows and erosion create inverted topography.

Pahoehoe texture on the surface of the Snow Canyon Overlook flow. The pencil points in the approximate direction of flow.

Columnar joints on the surface of the Snow Canyon Overlook flow

flow, you can see that the canyon's latest migration has already begun. The drainage is once again shifting west as surface water cuts down into the sandstone and carves a new canyon. We'll take a closer look at this process later in this vignette at stop 4, Hackberry Wash.

Now turn your attention to the rock beneath your feet. The Snow Canyon Overlook flow is thick, massive, and dark gray to black. The small, hollow spaces you see in this rock are air vesicles. Because gas bubbles tend to rise to the top of a liquid (think of the head on a glass of beer), these vesicles are concentrated near the surface of the flow. Note that the vesicles are elongate in shape, rather than round like typical air bubbles. The lava that made up this flow was very viscous. As it moved across the surface, it squeezed the air vesicles and pulled them along, so the elongation tells us the direction of flow. You may also notice that in some areas the surface of the rock shows a twisted, ropy texture. This type of surface texture, called pahoehoe (a Hawaiian term pronounced pa-HOY-hoy), forms as hot, fluid lava moves across the surface. Note how the ropy forms curve. The convex side of the curve points in the general direction of flow.

There are also deep cracks on the surface of the flow, called joints, some of which come together to form polygonal shapes. Joints that intersect to form polygonal shapes are called columnar joints because, if you look at a lava flow with such joints in a vertical section, such as a cliff face, the rock seems to be comprised of polygonal columns. Columnar joints form in the following way. As any material cools, the molecules within it slow down. Slower molecules take up less space than they did when they were warmer and moving faster, so some shrinking or contraction occurs. These joints form due to stresses associated with this contraction as the basalt cools. For an in-depth discussion of columnar jointing, see vignette 13.

For more insight into this lava landscape, let's move on to stop 2, where we'll look at the origins of the Santa Clara flow. From the parking area along Utah 18, look at the two small, dark hills, one just to the southeast, the other a little farther away and to the north. These are cinder cones that formed during the Santa Clara eruption. The fact that they're so well preserved tells us that these structures most likely formed during later stages of the eruption. Cinder cones formed early in an eruption are usually destroyed during later stages.

To better understand how these lava flows and cinder cones came to be, it helps to take a step back and look at the larger picture. Snow Canyon lies in the transition zone between two geologic provinces, the Colorado Plateau to the east and the Basin and Range to the west. In this area, tectonic forces are pulling apart and thinning the earth's crust, fracturing the rock. At times in the not-so-distant past, magma used faults and fractures in the Navajo sandstone as conduits to the surface. All magma bodies contain dissolved gases, referred to as volatiles. While the magma is deep underground, it's under very high pressure, so the gases remain dissolved. However, as the magma pushes its way to the surface, the pressure decreases and the volatiles are released. A can of soda is a good analogy for this process. The sealed can keeps the liquid within under pressure, so the carbon dioxide gas remains in solution. When you open the can, the pressure is released and some of the dissolved carbon dioxide is able to escape. The escaping gas causes the hiss you hear as you open the can. Since magma is much more viscous than water, its volatile gases can't escape as easily. As the gas bubbles move

A rounded volcanic bomb ejected from the cinder cone southeast of stop 2

toward the surface, they carry along bits of molten rock, which are ejected out of the vent when the gases reach the surface. These molten blobs cool and solidify in the air, falling back to the surface as cinders. Eventually, through repeated eruptions, a large mound, called a cinder cone, builds up around the vent.

Between the parking area and the cinder cone to the southeast lies a portion of the Santa Clara flow that erupted from vents near the base of that cone. Note the rugged, blocky texture of the flow's surface. This texture, with the Hawaiian name *aa* (pronounced AH-ah), forms in the following way: The upper portion of a viscous flow is exposed to air and cools relatively quickly. This upper crust insulates the underlying lava, which is still very hot and continues to flow. As it flows, it breaks apart the cooled crust above it, jumbling it up into the blocky texture you see here. A hike across an aa flow can be very damaging to your boots, your behind, and your pride.

Look at the various blocks that make up this flow. Some are relatively solid, but many are vesicular. The vesicles in this flow tend to be more rounded than those in the Snow Canyon Overlook flow, although some do show minor elongation. Also note the reddish brown color of many of the blocks. Basalt from the Santa Clara flow contains more iron than the other flows in the area. Oxidation of this iron forms iron oxide, or rust, which stains the surface of the rock, just as steel tools become rusty when left out in the rain. The rusty color of the Santa Clara flow helps differentiate it from other flows in the area.

Now carefully make your way across the basalt flow to the cinder cone. Note the range in size of the particles ejected from the vent. The vast majority are small, gravel-size cinders, but some are larger. Some of the midsize pieces are relatively smooth and rounded and have an aerodynamic look. As ejected pieces of molten rock spin and tumble in the air, they take on these shapes and are referred to as bombs. Also note that most of the cinders are chock-full of air vesicles. As we mentioned earlier, the action of escaping gases throws these cinders into the air, so it should be no surprise that the majority of them contain vesicles. Scoria is a term used to describe basalt that is extremely vesicular.

An unmaintained trail up the northwest side of the cinder cone leads to the crater at the top. If you choose to take it, be careful: the trail is steep and the cinders create a slippery surface. But the climb is rewarding, as it leads to a spectacular view. The trail ends at the crater rim on the north side of the cone. Looking north from this point, you can see the north cinder cone and the lobes, or fingers, of the flow that erupted from vents at its base. Now walk around to the south rim of the crater. From there you can see a ridge of white Navajo sandstone extending to the south. Look for a black area on the west slope of this ridge. This is a basalt dike that formed as lava erupted through a fracture in the sandstone. As mentioned above, the crust in this area is being pulled apart, causing joints in the Navajo sandstone to widen. Magma has used these joints as conduits to the surface in the past, and the dike in this ridge reveals the presence of one such conduit. Although these vents seem dormant now, the processes that led to their formation are still at work. Future eruptions cannot be ruled out.

Climb off the crater and proceed to stop 3, the trailhead for the Lava Flow Trail, where we'll continue our study of features of the Santa Clara flow. Blocks of basalt border the well-maintained trail. The vesicular nature and rusty color of many of the blocks are clues that you are indeed walking across the Santa Clara flow. Once again, take note of the rugged, blocky surface of this aa flow. In some areas, fractured linear mounds of basalt, known as pressure ridges, have formed. These ridges, usually several feet tall and 10 to 20 feet long, form as large, solidifying blocks of lava riding atop the molten material are jammed up against each other.

Just beyond the junction with the Whiterocks Trail, you'll see a deep hole on the north (right) side of the trail. Be careful around the

View north from a turtleback south of the Lava Flow Trail. The darker rock in the foreground is basalt from the Santa Clara flow. The lighter rock in the background is the Navajo sandstone. The dark area near the center is a collapse structure giving access to a lava tube. Note hiker (highlighted) *for scale.*

edge of this structure because the rocks aren't very stable. What you're looking down into is a lava tube. A lava tube forms when the surface of a flow cools and hardens to a solid crust, but molten lava continues to flow beneath it. When the eruption ceases, the lava drains out and a hollow tube remains.

Lava tubes aren't always easy to find. You may be walking across a basalt flow and not even know a lava tube is under your feet. Here in Snow Canyon, the process of erosion helped us locate this lava tube. Weathering at weak points in the tube's roof caused it to collapse, revealing the structure. The alignment of this collapse feature with others in this flow gives us a pretty good plot of the course of this lava tube.

Don't try to climb down into the tube here. For those who feel the need to explore, another collapse structure farther down the trail affords easier access. But before leaving this site, take a look at the small ridge of basalt west of the collapse feature. Here you can see another example of the ropy pahoehoe texture we first noted at Snow Canyon Overlook.

Proceed down the trail to a fork. The left fork goes to the West Canyon Overlook, and the right fork heads down into West Canyon. Take the right fork and keep your eyes on the north side of the trail. Within a couple hundred yards you'll see the opening of a much larger collapse structure. Carefully climb down into the opening. As you descend, you'll notice a distinct drop in air temperature, especially on a hot day. This is one of the perks of this hike. Along the walls you can see small, delicate drips, dribbles, and curtains that formed as molten rock dripped from the ceiling of the tube. Please leave these fine features intact for others to enjoy. When you're ready, follow the trail back to your vehicle and drive to stop 4, at Hackberry Wash.

At Hackberry Wash, you can see firsthand how this inverted topography came to be. From the pull-out, enter the wash to your left, turn right, and proceed about 400 feet, crossing the bike path. On your left, you'll see a 30-foot-high wall of basalt. On your right is a wall of sandstone. The basalt on the left is still jagged, showing few signs of erosion. The sandstone wall, on the other hand, is smooth and rounded. As discussed before, the erosion-resistant basalt of the Santa Clara flow filled the previous drainage, and water running off the rocks above is now cutting down into the sandstone along the edge of this flow. If you examine some of the sand in the wash, you'll see that it's almost completely composed of light-colored quartz grains eroded from the sandstone. Darker grains of basalt make up only a small percentage of the total, telling us that much more sandstone has been eroded than basalt. This same process is occurring along the western edge of the Santa Clara flow, where Snow Canyon is once again migrating west. The cycle continues.

The delicate, drippy texture of the walls of the lava tube
gives it an almost surreal feel. Pocketknife for scale.

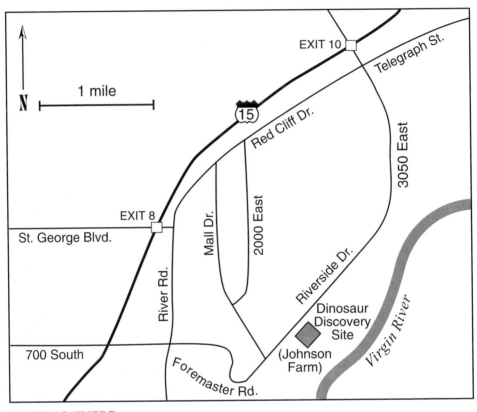

GETTING THERE

From exit 8 on I-15 in St. George, drive 0.1 mile east to River Road and turn right. Drive 1 mile south on River Road to Foremaster Road, then turn left (east) for 1.7 miles on Foremaster Road, which becomes Riverside Drive, to the signed turnoff on the right to the St. George Dinosaur Discovery Site at Johnson Farm. The facility is visible from Riverside Drive.

2 WHERE DINOSAURS ROAMED
St. George Dinosaur Discovery Site at Johnson Farm

The St. George Dinosaur Discovery Site at Johnson Farm sits on a rocky bluff within the city of St. George. While clearing land in February 2000, St. George resident Sheldon Johnson uncovered block after block of sandstone and mudstone containing some of the best-preserved dinosaur prints in the world. The fossils reveal not only footprints of these great beasts, but also impressions of their claws and skin.

The large blocks of stone at the site are from the Moenave formation, which dates to about 200 million years ago, in early Jurassic time. This rock unit, which contains sandstones, mudstones, and shale, rests upon the late Triassic Chinle formation, visible in the nearby Virgin anticline (discussed in vignette 6) and underlies the Jurassic Kayenta formation, seen in the prominent sandstone cliffs surrounding St. George. We'll talk about the environment in which these deposits were laid down in just a bit, but first let's focus on geologic time and dinosaurs.

The first scientists to study dinosaurs looked in awe at the huge bones they had unearthed and assigned the long-dead creatures a name that suited their fearsome-looking teeth and sharp claws. In 1842, British paleontologist Richard Owen coined the term *dinosaur*, meaning "terrible lizard." Of course, we now know that dinosaurs were not really lizards at all.

The first evidence of dinosaurs appears in the rock record in earliest Mesozoic time, about 245 million years ago. Dinosaurs dominated the terrestrial environment for 180 million years before their mass extinction at the end of Mesozoic time, about 65 million years ago. Scientists generally agree that a catastrophic event, perhaps a huge meteorite impact, killed off most of the life on our planet, including the dinosaurs.

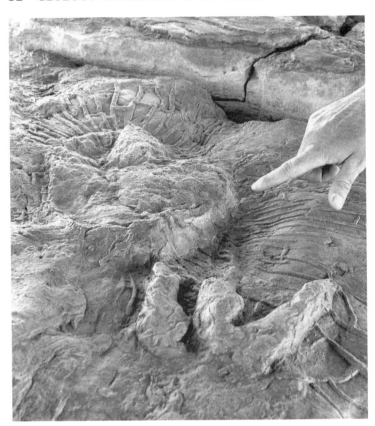

This three-toed track is deeply pressed into the substrate. Fractures radiate outward, attesting to the weight of the animal that strode here.

Dinosaur bones fill collections in museums around the world and give us important information about dinosaurs. Bones allow us to reconstruct these animals and make hypotheses about their physical structure and appearance. There is, however, a specialized branch of paleontology that focuses not on the physical remains of extinct organisms, but on other clues they leave behind, known as trace fossils. This field is called paleoichnology, which means the study (*ology*) of ancient (*paleo*) footprints (*ichnos*). Despite its name, there's more to paleoichnology than footprints. Dinosaur paleoichnologists look at fossilized tracks, nests, skin imprints, feeding traces, gastroliths (stomach stones), vomitus, and even coprolites. (Coprolites are fossilized poop. Yes, some scientists specialize in the study of fossilized poop!) Trace fossils allow researchers to delve into paleoethology, the study of the behavior of extinct animals. Tracks and trackways (multiple tracks) help us understand how dinosaurs walked and ran; nests give clues about familial relationships; and petrified vomit and coprolites give us a glimpse into eating habits and digestive functions.

Swim tracks are made as a dinosaur swims across shallow water.
Each stroke produces marks as claws scrape bottom.

The site at Johnson Farm offers an assortment of trace fossils, including footprints, skin imprints, and swimming tracks. In addition, a variety of features accompanying the trace fossils allow us to hypothesize about the environment in which these dinosaurs roamed. Browsing the many footprints on display here, you'll see that some of them form concave impressions in the rock, while others protrude from rock faces. Concave impressions can be primary prints or ghost prints. Primary prints are made in the surface on which a dinosaur walked; the surface you see was actually in contact with a foot. Ghost prints form in layers beneath the surface. If sediment is soft enough, surface impressions can be carried down into the substrate, where they form roughly similar prints in buried layers. Though ghost prints may be geometrically similar to primary prints, they typically lack the detail found in surface impressions. Impressions are sometimes called molds.

In contrast, many of the prints on display at Johnson Farm are convex features called casts. In this type of trace fossil, after the prints were made, sand swept over the area and filled the molds. Over time, this sand

A close look at a claw impression

cemented together to form sandstone. When humans overturned and separated these blocks of rock 200 million years later, the lower mudstone preserved molds and the upper sandstone preserved casts. Some of the most impressive footprints at this site are large casts so detailed that you can see a claw mark at the end of each big toe.

The sediment that later became the mudstones and sandstones of the Moenave formation was deposited near sea level when this landmass was closer to the equator than it is today. Streams in that generally arid landscape deposited coarse sediment in their channels, while finer sediment ended up in the shallow lakes into which these streams flowed. At Johnson Farm, the rocks bearing dinosaur prints have characteristics pointing toward origins in the latter, a freshwater lake environment. If you take a close look at the blocks of stone around you, you'll see many beautifully preserved symmetrical ripple marks. Ripple marks are found on sand dunes, shorelines, and stream channel bottoms. Where flow is unidirectional, as in streams, the ripples are asymmetrical, with a gradually sloping upstream face and a steeper downstream face. On the shoreline of a lake, though, water moves in with breaking waves and back out with wash, creating ripple marks with evenly sloping faces. While the latter predominate at Johnson Farm, some blocks at the site do have asymmetrical ripple marks. The combination tells us that stream channels once carried water down into a lake at this site.

Many of the prints at Johnson Farm are found on surfaces with extensive mud cracks, and some prints have mud cracks radiating outward from them. Mud cracks form when lakes dry up. Lake sediment has very fine-grained clay minerals that absorb water and swell. When a lake recedes, once-saturated sediment is exposed to the atmosphere and dries out. The clay minerals release water and shrink, producing a pattern of polygonal mud cracks on the desiccating surface. The extensive mud cracks in the rocks at Johnson Farm suggest that the trackways here were made by dinosaurs that were eating and drinking at the edge of the receding lake—perhaps a vital source of water in an arid climate.

What makes us think the climate was arid here during early Jurassic time? Seasonal swelling and shrinking is commonplace in desert lakes. In the spring, such lakes often expand as they receive rainfall and snowmelt from surrounding highlands, but in the summer, they rapidly evaporate, leaving cracked mud where water once lay. Also, desert lakes tend to be quite salty, and when such lakes desiccate, salt crystallizes within their sediment. Salt casts in the Moenave formation attest to this sort of environment.

Fossils called rhizoconcretions provide further evidence of aridity. Desert soils are characterized by abundant calcium carbonate, which forms caliche or hardpan, a white, concretelike layer below the surface. As rainfall infiltrates these soils, it becomes saturated with dissolved calcium carbonate. Desert plants take up this water from soil but leave precipitated calcium carbonate behind as expanding, hard layers that surround their root systems. When these plants die, the calcium carbonate impressions of their roots often remain in the soil; these impressions are called root casts. In the root casts at Johnson Farm, silica has replaced the calcium carbonate over time. Since silica is much harder than calcium carbonate,

Rhizoconcretions indicate an arid environment and provide evidence of plant life during the time dinosaurs lived here.

the Johnson Farm rhizoconcretions are very durable, a
requirement for preservation
over the 200 million years since
these plants lived.

Trace fossils are most
commonly classified by their
association with one of the
following five activities: resting
or hiding, dwelling, feeding,
grazing, or movement. The
footprints at Johnson Farm obviously fall into the movement
category. Paleoichnologists
assign footprints and other
trace fossils names based solely
on morphologic criteria—that

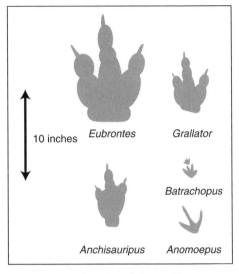

Prints found at Johnson Farm

is, based on the form and structure of the fossil. They don't include
subjective interpretations about the originator of the trace since, in most
cases, it's impossible to determine exactly which organism made the trace
fossil. Only in extremely rare instances have body fossils been found in
direct association with trace fossils. This doesn't mean paleontologists
don't try to figure out which dinosaur made which print. They do, but
they use different names for tracks than for dinosaurs because of the
uncertainty in correlating prints with print makers.

Let's take a look at the Johnson Farm prints. There are at least five
distinct types of prints, all of which have been tentatively identified. The
largest tracks at Johnson Farm, three-toed prints more than 10 inches in
length, are assigned the name *Eubrontes*. We believe the actual bipedal
dinosaur that made these prints may be from the genus *Dilophosaurus*
and that it was probably a meat eater. Larger individuals probably stood
7 feet tall at the hip, stretched 20 feet from nose to tail, and weighed
over 1,000 pounds. Other creatures that made tracks called *Grallator* and
Batrachopus also strode the desiccating shorelines at what is now Johnson
Farm. The dinosaur that made *Grallator* was a small, fast predator like
the dinosaur *Coelophysis*, the fossilized bones of which have been found
at Ghost Ranch, New Mexico. The animal that made *Batrachopus* was
not actually a dinosaur, but rather a terrestrial crocodile, although very
different from the crocodiles of today. Here its tracks include impressions of smaller front feet and larger rear feet. Two other sets of tracks
have been tentatively identified as *Anchisauripus*, three-toed prints 4 to 7
inches long, and *Anomoepus*, delicate prints only 2 inches long.

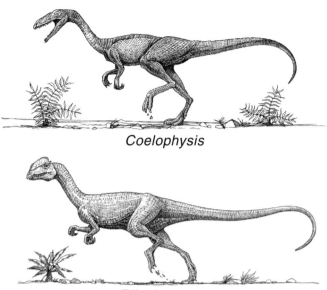

Pen-and-ink drawings of two dinosaurs that may have made tracks at Johnson Farm —artist, David Smee

Coelophysis

Dilophosaurus

Johnson Farm is an ideal place to see a variety of dinosaur prints in a sheltered setting. It affords a rare opportunity to appreciate some of the best dinosaur tracks in the fossil record. In addition, characteristics of the rock allow us to visualize the dinosaurs in their ancient environment, roaming the edge of a desert lake in search of food or water. Most of the dinosaurs that made this trackway were upright walkers of small or moderate size. For the adventurous traveler undaunted by dirt roads and desert heat, other dinosaur prints in the Moenave formation can be found in Warner Valley, about 10 miles west of St. George; ask for directions at the local BLM office. To see some really big footprints, visit the Copper Ridge trackway near Moab, discussed in vignette 31, where you can stand by the footprints of a great sauropod, one of the true giants of Mesozoic time.

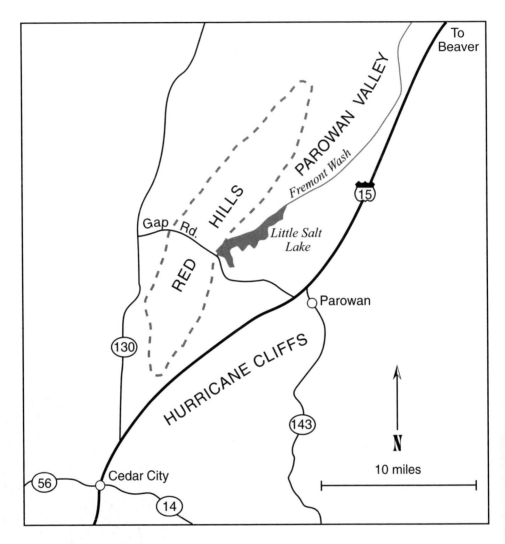

GETTING THERE

From Cedar City, drive north on I-15 to exit 62 and Utah 130. Drive 14 miles north on Utah 130 to Gap Road, then turn right (east) for 0.8 mile on Gap Road to the west edge of Parowan Gap. Gap Road takes you across the Red Hills to Little Salt Lake east of Parowan Gap.

3 WHEN WATER CLEAVES ROCK
Parowan Gap

A water gap is a streamcut valley that crosses a ridge, and all water gaps pose an interesting question: How did the stream seemingly flow up one side of a ridge and down the other to incise a gap? Here at Parowan Gap, a water gap without a present-day stream, there are two possibilities involving two different types of streams. Let's begin our discussion at the gap's western edge. Parowan Gap cuts across the Red Hills, a landform composed of Jurassic, Cretaceous, and Tertiary sedimentary units

The prominent notch at the west end of Parowan Gap

19

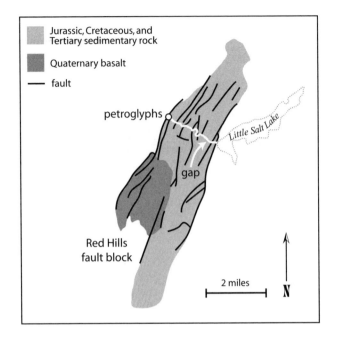

Jurassic, Cretaceous, and
Tertiary sedimentary rock

Quaternary basalt

— fault

petroglyphs

Little Salt Lake

gap

Red Hills
fault block

2 miles

N

Geology of Parowan Gap

as well as black Quaternary basalt in the southwest. Many people think that the gap is just the more prominent V-shaped notch at the western edge of the Red Hills, but the gap actually transects the entire width of the Red Hills, from here to Little Salt Lake on the east.

A superimposed stream may have cut this gap. Superimposed streams form on sediment that has filled a basin and completely submerged any preexisting topography. Suppose the Red Hills existed as a ridge 2 million years ago. Since that time, enough sediment may have been shed from the uplifted Colorado Plateau to fill the Parowan Valley and cover the Red Hills. Streams would have flowed down a gentle slope from highlands in the west to a low valley in the east. But streams erode and transport sediment, so over time they would have gradually deepened and broadened their valleys. Downward erosion would have been gradual, and the stream crossing the Red Hills would have maintained its course even as it cut downward beyond the sediment and into the exhumed Red Hills ridge. The other possibility is that an antecedent stream formed the gap. An antecedent stream is one that continues to flow along its established course when faulting causes uplift in its path. The stream can continue to cut a deepening path in the fault block as long as its erosional ability keeps pace with uplift. If the rate of uplift is greater than the stream's erosional power, the fault block becomes an impediment to flow. The stream then either flows around the block or ponds against it. So which type of stream formed Parowan Gap?

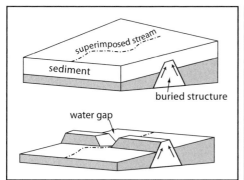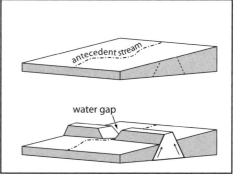

Water gaps created by superimposed (left) *and antecedent* (right) *streams*

If you entered the water gap from Utah 130 on the west, drive east through the gap to Little Salt Lake and then return to the western edge, paying attention to the rock walls around you. If you got here from the east, you've already seen these rock units. Some beds in the rock dip down to the east, some dip west, and some are more or less horizontal. We know that sediment is initially deposited in flat-lying beds; geologists refer to this as the principle of original horizontality. When we see a variety of orientations in rock units, such as here in the Red Hills, we know something happened to disturb that original horizontality. Here at Parowan Gap, faulting caused the disruption. Many faults, with a dominant north-south orientation, crisscross the Red Hills. Faulting and fracturing here are the result of internal stress within the uplifted Red Hills fault block. (A fault block is loosely defined as rocks that move together in the same direction.) Since faulting is present here, it's tempting to go with the hypothesis of water gap formation by an antecedent stream rather than a superimposed stream. However, it could be that these faults are old features predating the more recent stream incision that formed the gap. So let's look for other clues.

Little Salt Lake lies against the eastern margin of the Red Hills. It's really not a lake but rather a playa, a remnant of a freshwater lake that has dried up. The white color of the playa bed comes from salt that precipitated during the most recent stage of desiccation, which probably occurred many thousands of years ago. Though there's sometimes shallow water here in early spring from snowmelt above the Hurricane Cliffs, most of the time the playa is dry. Most playas in the southwestern United States are remnants of large lakes that reached their maximum size about 14,000 years ago when the climate was wetter and cooler (for more on this topic, see vignette 5 on Sevier Lake). This may have been the case for Little Salt Lake. The water gap tells us that the stream flowing into

Some bedrock within the fault block is horizontal.

Other bedrock outcrops are tilted at near-vertical angles due to interior faults that break the Parowan fault block into numerous smaller blocks.

Little Salt Lake once cut across the Red Hills on its westward journey into the Great Basin. The playa tells us that this flow stopped when the Red Hills rose up in the stream's path. Logic tells us that the presence of Little Salt Lake supports the hypothesis of antecedence, and that upward movement of the Red Hills fault block eventually accelerated beyond the stream's ability to cut down. At that point, water began filling the basin east of the Red Hills to form Little Salt Lake. When the climate became warmer and drier, the lake dried up, but not without leaving the salty playa behind as evidence of its existence.

Parowan Gap is a dramatic feature, especially here at its western edge. Modern humans are not the only ones who have found it compelling. The Jurassic rock exposed here is decorated with petroglyphs of unknown age and origin, apparently of several different cultural styles. Though the origins of the petroglyphs are uncertain, it's clear that this

*The Little Salt Lake playa is a remnant of the lake created when
the rising Red Hills fault block outpaced stream incision.*

place was important enough to have served as a canvas for prehistoric artists. The Zipper Glyph, one of many interesting images in the water gap, is believed to be a calendar. It has 180 notches cut across a central, looping line, one notch for each of the 180 days it takes for the sun to move across the horizon between the summer and winter solstices. The orientation of ancient stone cairns found near the gap also seems to be related to the solstices. The human need to understand time and space is reflected in these artifacts, and probably in others, too, if only we were clever enough to divine their purposes. The petroglyphs here are significant enough to be listed in the National Register of Historic Places, so enjoy looking at them but, as with all rock art, please do not touch.

The Zipper Glyph is thought to be a calendar. A shadow from a nearby ridge follows the notched line along an annual cycle.

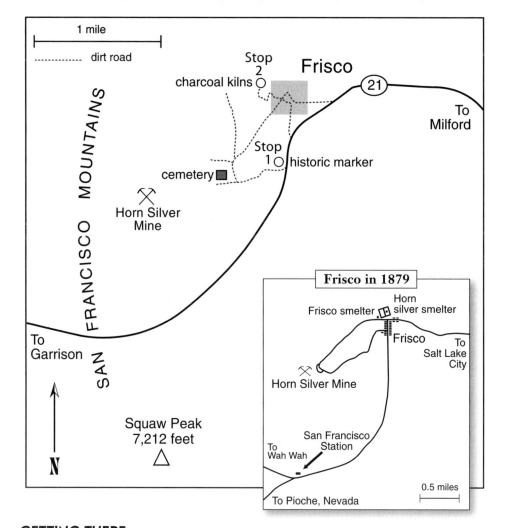

The following labels appear on the map:

1 mile
dirt road
Stop 2
charcoal kilns
Frisco
21
To Milford
Stop 1
historic marker
cemetery
Horn Silver Mine
SAN FRANCISCO MOUNTAINS
To Garrison
Squaw Peak
7,212 feet
N

Frisco in 1879
Frisco smelter
Horn silver smelter
Frisco
To Salt Lake City
Horn Silver Mine
San Francisco Station
To Wah Wah
To Pioche, Nevada
0.5 miles

GETTING THERE

To get to the remnants of the town of Frisco and the Horn Silver Mine, go 46 miles west of downtown Beaver on Utah 21 through the towns of Minersville and Milford. After Milford, look for the bed of the old Utah Southern Railroad line to Frisco, sometimes visible on the north side of the road. Stop 1 is about 14 miles beyond Milford on the right. A stone historic marker in front of a sheltered picnic area bears a plaque noting the mine's location and giving a brief history of the mine and town. Behind the picnic area, the dirt road forks; about 0.4 mile down the left branch is the Frisco cemetery and a view of the mine works. Stop 2 is Frisco proper and its charcoal kilns. To get there, go left from the picnic area on Utah 21 back toward Milford for 0.2 mile and follow an unmarked access road on your left for about 0.2 mile. You can also park at the picnic area and walk to both stops. The kilns, listed in the National Register of Historic Places, are in the northwest section of the town remnants. For your own safety and the preservation of what is left of Frisco, please respect fencing, gates, and No Trespassing signs in the area and be careful around structures.

4 AN ALLURING LUSTER
Frisco and the Horn Silver Mine

A thousand men, say, go looking for gold. After six months, one of them's lucky. . . . His find represents not only his own labor, but that of 999 others to boot. That's six thousand months, five hundred years, scrambling over a mountain, going hungry and thirsty. An ounce of gold is worth what it is, mister, because of the human labor that went into the finding and getting of it.

Howard as played by Walter Huston in
The Treasure of the Sierra Madre

The history of Frisco, Utah, and the Horn Silver Mine is the story of much of the West in the latter half of the nineteenth century. It's a classic tale well dramatized in the 1948 movie *The Treasure of the Sierra Madre*, directed by John Huston. Key elements of this tale usually include mine speculation, mineral discovery, sudden wealth, railroad expansion, lawlessness, and, in the end, a bust. While the story may have been repeated throughout the West, each mining town and companion mine also had its unique triumphs and tragedies. Wandering through Frisco's ghostly remains and the nearby mining structures is a good way to sense its own particular story. Beehive kilns that produced charcoal for the smelting process are well preserved. Scattered shards of rusted metal and opalescent glass lie among the buildings, testifying to a once-thriving human landscape. What drew people here? Why did they leave? The answers trace back to a hydrothermal process that occurred here, creating high concentrations of one of our most valuable metals: silver.

Frisco lies east of the San Francisco Mountains, which are part of the Wah Wah and Tushar mineral belt. Mineral belts are typically associated with past volcanism, and volcanoes were indeed active in this area for 25 to 30 million years during Oligocene and Miocene time, about 36 to 5 million years ago. During that time, the earth's surface stretched and thinned, allowing magma to reach the surface. The San Francisco Mountains themselves are intrusive rock—a type of rock formed from magma that injects itself into surrounding rock beneath the earth's surface. The density and viscosity of the magma determines the extent of the intrusion. (Vignette 32 discusses another intrusive range, the La Sal Mountains.) Millions of years of uplift and erosion have uncovered and exposed the intrusion that makes up the San Francisco Mountains.

During the initial intrusion process, cooling magma released hydrothermal fluids—hot water laden with minerals. These fluids followed and filled cracks and cavities in the surrounding rock. Veins of mineral deposits formed as the fluids cooled and their minerals crystallized. Following a fault, the hydrothermal fluids under the Frisco area flowed into an existing zone of weakness and deposited an ore body—a deposit of valuable, extractable minerals—between limestone on one side and volcanic rock on the other. Millions of years later, miners at Frisco constructed mine shafts through two classes, or grades, of ore.

The view south to Squaw Peak

The first class was high quality and lead-free. The second class—the majority of the ore here—contained 30 to 60 percent lead. Each class required smelting to extract the minerals. The main deposit, 40 to 60 feet wide at the discovery location and wider below, extended no deeper than approximately 1,000 feet, with pockets of zinc and copper ore at depths of 700 to 800 feet.

Silver has a whitish luster but tarnishes to black. Despite the tarnish, humans throughout history have been attracted to this metal for both its appearance and its functionality. Ancient slag dumps in Asia Minor and on islands in the Aegean Sea attest to early discoveries of how to separate silver from lead. Archaeologists have found silver ornaments and vases dating back to ancient Troy and Egypt. In the sixth century BC, the Athenians elevated silver to precious-metal status and were one of the first civilizations to mint silver coins. Silver was the basis for most currency until the nineteenth century, when gold replaced it. Because of silver's ability to conduct heat and electricity, the ease with which it can be shaped, and other properties, it's still important in the manufacture of utensils and electronics and in dentistry and photography.

In June 1849, a little over a year after the discovery of gold at Sutter's Mill in California, the first silver nugget was removed from the Comstock Lode in Virginia City, Nevada. Eventually miners would extract $400 million in minerals from that lode. While Mormons participated in early mineral discoveries in California and Nevada, Mormon leader Brigham Young initially discouraged mining in Utah. He felt it would divert

This boulder of metal ore sits near the Frisco cemetery.

manpower from the agricultural needs of the Mormon community and encourage non-Mormon settlement. However, recognizing that a strong economy needs mining and its products, especially coal and iron, he later changed his stance.

In 1863, commercial mining officially began in Utah, urged on by the U.S. Army. Gathering wealth was not the only motivation. The year before, concerned that the Mormon community might align with Native Americans or establish an independent commonwealth, President Lincoln had directed Col. Patrick E. Connor and 750 California and Nevada volunteer soldiers to establish Fort Douglas in the Salt Lake Valley, guard Overland Mail routes against Indian attacks, and also "keep an eye on Brigham." Connor arrived in Utah with a strong bias against Mormons. His first wish was to establish martial law in the territory, and to this end, he requested three thousand additional troops. With the Civil War on, however, the army was reluctant to start a war with Mormon settlers. Instead, Connor was instructed to pursue peaceful alternatives to the "Mormon question." He decided the best solution was to dilute Mormon influence by luring large populations of non-Mormons into the area with the promise of mineral wealth. Troops in the territory were allowed to prospect in the Wasatch and Oquirrh Mountains; Connor and his men ultimately formed a mining company with claims in Bingham Canyon, Little Cottonwood Canyon, and Tooele County. Early mining was an expensive enterprise that usually required the shipping of ore for smelting and refinement (more about these processes in a moment). However, the completion of the transcontinental railroad at Promontory, Utah, in May 1869 eased shipping costs and encouraged the growth of many mining operations. By 1875, prospecting and mining were established industries in Utah.

The San Francisco Mining District, created in 1871, encompassed seven square miles, including the San Francisco Mountains. Two prospectors, James Ryan and Samuel Hawkes, had been mining the mineral galena, which often contains silver, in this area. On September 24, 1875, they landed a pick in an outcrop they had regularly passed and discovered an ore body. Folklore has it that the ore Ryan and Hawkes discovered was so rich you could whittle slivers from it that curled in the shape of mountain sheep horns—hence the name, the Horn Silver Mine. Ryan and Hawkes staked their claim, then sold it in February 1876 for $25,000, thinking the ore vein was limited.

Under new ownership, the Horn Silver Mine came to prosperity. The main shaft, located about 200 feet from the outcrop Ryan and Hawkes discovered, followed the vein in its original fault to a depth of 255 feet, with five horizontal levels. After two years of digging, the shaft reached a depth of 400 feet. In 1879, banker Jay Cooke bought the property for $5 million and created the Horn Silver Mining Company of Utah. That same year, the *United States Annual Mining Review and Stock Ledger* called the Horn "unquestionably the richest silver mine in the world now being worked." By that time, the mine had produced 25,000 tons of silver-laden ore. In 1880, the Union Pacific Railroad extended the Utah Southern Railroad to Frisco, facilitating Frisco's growth and supporting mine operations. You can still see the railroad bed today (see "Getting There").

In 1877, in a reflection of the Horn's growing success, two smelters and five beehive charcoal kilns were constructed at Frisco to aid in the silver processing. The kilns at Frisco, made of granite with lime mortar, are captivating parabolic-dome structures. Their diameters vary from 16 to 26 feet, with walls 18 to 30 inches thick at the base. To create charcoal in the kilns, piñon wood, purchased at $1.25 per cord in 1880, was burned for three to seven days and cooled for another three to six. Prior to construction of the kilns, charcoal was made in pits, a process that took fifteen to twenty days. Thus, the efficiency of the kilns offset their $500 to $1,000 construction costs. Two doors provided access to each kiln for adding wood and removing ash, and three rows of vents near the bottom enabled workers to regulate the fire. Forty-five cords of wood could be processed at one time in the larger kilns.

Each smelter consisted of several furnaces with hearths made of volcanic tufa. Since water, needed for cooling components, was scarce, sometimes only one furnace ran at a time. Smelting is the process of separating impurities from metal by melting and oxidation. In Frisco, workers roasted the ore over wood or low-grade charcoal to start the oxidation process and break down the ore prior to smelting. The ore was then placed in the furnaces for melting and further oxidation. Charcoal created in the Frisco kilns provided a high-intensity heat that melted the ore and produced carbon. Carbon reduces the amount of oxygen in ore by combining with it to produce carbon dioxide, a gaseous by-product. Finally, workers added an iron oxide flux, a material that bonds with lightweight waste materials and makes them easy to skim away, leaving only silver and other valuable metals. For each ton of high-grade ore processed from the Horn, 75 to 200 ounces of silver were extracted. Lesser-grade ores produced 30 to 75 ounces. The kilns are the only intact remnants of this process.

The 1880 railroad extension to Frisco not only facilitated shipping ore out, it also eased transportation of workers, service providers, and families in. The town of Frisco, where these immigrants lived and all too often died, developed about 1 mile northeast of the Horn's main shaft. From

Charcoal kilns, with their characteristic beehive shape

eight hundred citizens in 1880, Frisco's population peaked at between five and six thousand just a few years later. The town boasted a post office, hospital, school, stores, and other amenities, along with a telegraph line to Beaver. During its heyday, Frisco purportedly had twenty-one saloons, along with numerous brothels and gambling dens. One reporter called the town "Dodge City, Tombstone, Sodom and Gomorrah all rolled into one." Daily killings eventually prompted the town fathers to hire a "town tamer," Marshal James Pearson from Pioche, Nevada, who arrived with four policies: Frisco would have no jail; Pearson would make no arrests; there would be no bail or appeal of his order; and outlaws had two choices, "to get shot or get out." The marshal reportedly killed six men on his first day on the job.

The town prospered with the mine. Through early 1885, an estimated $20 million worth of minerals, some 204,607 tons, were mined. Then, on February 12, 1885—fortunately, just after the night shift had surfaced—ground-shaking tremors signaled a cave-in. In their haste to gather wealth, mine operators had timbered the mine inadequately, and the timber supports could no longer support the rapid pace of ore extraction. In addition, recent rain and snow had soaked the ground, adding weight. The collapse was so cataclysmic it broke windows 15 miles away in Milford and effectively closed access to the richest part of the ore body.

Mining operations resumed after the cave-in. A new shaft 200 feet east of the original shaft eventually reached 1,600 feet deep. Still, the cave-in was the beginning of the end for Frisco. Fluctuating mine operations and moving smelting operations to the Salt Lake Valley, where smelters used less-expensive coke, reduced the need for town services. By the late 1920s, only one hundred residents remained and the mine manager doubled as postmaster. Sporadic mining continued until the 1950s, when it ceased on any scale of importance.

The Frisco cemetery at stop 1 is a sobering reminder of the hardships of life in a mining town. Many of the graves belong to infants and toddlers, reflecting high child mortality rates on the frontier. The epitaphs record parents' anguish. "The self same hand from whence 'twas given, Has taken back our babe to heaven," is inscribed for Henry J. Barrett, who died days before his first birthday. Tommy James's presence here was even briefer. His parents remembered him: "Whose all of life's a rosy ray, Blushed into dawn and passed away." Why live in such an unforgiving location, in a place that steals the lives of those most dear to you? The dream of mineral wealth, a constant throughout much of human existence, has an alluring power.

Headworks at the Horn Silver Mine

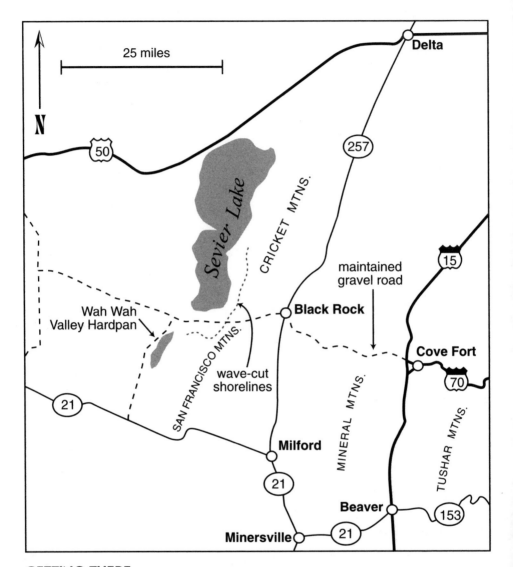

GETTING THERE

From Cedar City, drive 50 miles north on I-15. At exit 109, go 30 miles west on Utah 21 to Milford, then turn right (north) on Utah 257 for 23 miles to the tiny community of Black Rock. Turn left (west) onto the maintained gravel road at Black Rock. Drive about 12 miles to a low rise that reveals the Sevier Lake Basin to the northwest. Park on the side of the road. During a wet spring, the basin may hold shallow water, but in summer, it appears bright white from sunlight reflecting off the salt deposits.

5 A DESICCATED REMNANT OF LAKE BONNEVILLE
Sevier Lake

You are standing on the shoreline of a huge lake that stretches north as far as the eye can see. The blue water sparkles in the sun as shorebirds dip their long beaks into the shallow water in search of food. What? You say you can't see the lake? Well, that's because you have to look not only across space, but also across time. Sevier (pronounced suh-VEER) Lake is currently an arid desert basin, but 15,000 years ago it was part of a giant lake that covered much of Utah and extended north into Idaho and west into Nevada.

Anybody who has visited southern Utah understands that water is scarce here. But water was once quite plentiful in this region. During Quaternary time—the last 1.8 million years of geologic history—many currently dry valleys in the West filled with water to form pluvial lakes, bodies of water that shrink and swell with changing climate. Utah had the largest pluvial lake of all. The vast dry valley that lies before you is a remnant of pluvial Lake Bonneville. Two subbasins of Lake Bonneville still hold water: Great Salt Lake, northwest of Salt Lake City, and Utah Lake, south of the city. True to its pluvial nature, Lake Bonneville has changed with climate, achieving its maximum size as ice sheets swept over continents and smaller glaciers filled mountain valleys during the last ice age. When the climate shifted to the hotter, drier conditions we continue to experience today, the lake shrank to form disconnected lakes and immense dry flats like this one, also known as playas.

You are standing in one of the southernmost basins that filled with water and joined with others to form Lake Bonneville. Several clues point to the former existence of a large lake in this basin. If you

Panoramic view (top left and right) *of Sevier Lake playa, one of the southernmost basins of Pleistocene Lake Bonneville*

look north of the road along the western flank of the Cricket Mountains and south of the road at the San Francisco Mountains, you can see a prominent bench set against the bordering highlands. This is the old Bonneville shoreline, which sits at an elevation of 5,090 feet above sea level. Carved into rock and sediment by breaking waves, such terraces provide important evidence for determining the spatial extent of ancient lake systems. Waves represent the movement of energy imparted to water by blowing winds, and a large body of water produces proportionally large waves. When waves break against the shore and release that energy, they erode a flat surface called a wave-cut terrace. Such terraces often butt up against steep cliffs that represent the limit of wave erosion. Terraces form during times of relative stability when a lake's level remains constant. The longer a lake remains at the same elevation, the flatter and broader the corresponding terrace will be. Wave erosion is a product of wave impact (the force of breaking waves) and abrasion (the grinding action that rounds gravel, cobbles, and fragments of shell on beaches). If you stroll along this high bench, you'll find numerous boulders and cobbles with smoothed surfaces, evidence of waves that broke upon this rocky shore long ago.

Determining the maximum size of Lake Bonneville would seem to be straightforward: simply determine the land surface below the highest wave-cut terrace elevation within the lake's connected subbasins. However, there's a complicating factor: the highest terraces in the central basin are higher in elevation than the highest terraces in more distant basins. This results from an interaction between the lithosphere and the semi-molten asthenosphere, which together exist in a floating balance

View north onto a terrace cut by Lake Bonneville waves

The wave-cut terrace continues to the south.

Gravel accumulation marks geologically recent shorelines of the receding Sevier Lake.

called isostasy. Increasing the weight of the lithosphere depresses it into the asthenosphere, just as loading more people into a boat causes it to ride lower in the water. A body of water as immense as former Lake Bonneville adds a huge amount of weight to the surface, especially in its central basin, where it's deepest. This depresses the lithosphere. When a pluvial lake dries up, the surface responds by rebounding, but it does so very slowly. The greatest rebound occurs where the lake was heaviest, typically at its center. Hence, Lake Bonneville's central shorelines surrounding mountainous islands are higher than those in distant reaches of the ancient lake.

By connecting high shorelines at the margins of the lake, scientists have estimated that at its greatest extent, Lake Bonneville covered 20,000 square miles—more than the combined surface areas of modern-day Great Lakes Erie and Ontario. Bonneville was similar in surface area to Lake Huron (23,000 square miles) and Lake Michigan (22,300 square miles). With a maximum depth of over 1,000 feet, Lake Bonneville was deeper than all of the Great Lakes except Lake Superior.

Researchers have determined the history of Lake Bonneville's rise and fall from sediment cores and historic lake shorelines. Lake Bonneville grew quickly during the last glacial stage, which peaked about 20,000

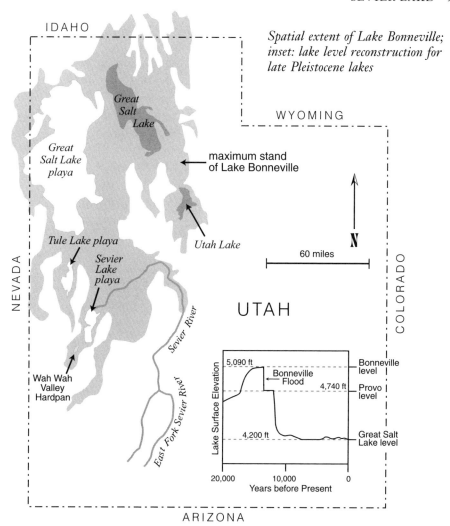

Spatial extent of Lake Bonneville; inset: lake level reconstruction for late Pleistocene lakes

years ago. Continental ice sheets covered much of Canada and flowed south into the northern United States at that time. As the glaciers receded, meltwater fed the rising lake until it reached its maximum surface elevation of 5,090 feet above sea level about 15,000 years ago. Since a drainage divide at that same elevation in southern Idaho provided an outlet for the lake, continued growth was impossible. As more water flowed into the giant lake, it just spilled over the divide. But all of that water began to have an impact on the spillway, soon leading to the Bonneville Flood, a sudden erosional episode that broke through the divide. Floodwaters surged north into the Snake River, flowed west across southern Idaho, north along the Idaho-Oregon border, west across southern Washington, and into the Columbia River near the town of Pasco. From there, the flood wave hurtled west along the Washington-Oregon border and finally spilled into the Pacific Ocean.

The volume of water released in this event undoubtedly left a long trail of death and destruction in its wake. The flood removed 350 vertical feet from the pass (the new, lower divide is today's Red Rock Pass in Idaho) and reduced the lake's surface area from 20,000 square miles to 14,400 square miles. Scientists called this smaller body of water Lake Provo. About 12,000 years ago, Lake Provo dwindled rapidly as the climate changed. By the onset of Holocene time, 10,000 years ago, the surface of the renamed Lake Gilbert was only about 50 feet above the level of present-day Great Salt Lake. The lake rose and fell by only small amounts during the ensuing 10,000 years, eventually becoming the Great Salt Lake we're familiar with today.

Why did Lake Bonneville grow so large? The answer probably lies in global wind patterns. The jet stream, a zone of high-velocity wind, carries moist air from the Pacific Ocean into the northwestern United States and Canada. This pattern of wind motion has a strong influence on the climate in coastal Washington and Oregon, making it obviously much different from that in coastal areas just a few hundred miles south. During the periods when ice sheets covered all of Canada and parts of the northern United States, the topographic and climatic influences of the very thick ice sheet split the jet stream. The temperature of the ice sheet also affected wind motion, establishing a strong region of high pressure, called an anticyclone, over the glaciated area. This anticyclone drove part of the moisture-laden jet stream north and the other part south into the previously arid Great Basin. This is a good example of feedback in a natural system. Climate change caused continental glaciation, which in turn changed the path of the jet stream, which led to further changes in climate.

Much of western Utah sits in the Great Basin, an inwardly draining region of the southwestern United States. Rain that falls here never makes it to the sea; instead, it collects in the many closed basins that make up the Great Basin. Evaporation is one of the few mechanisms by which water escapes this closed system. This helps explain why the Sevier Lake playa is a salt flat devoid of vegetation. Playas develop in pluvial lake basins that were once filled with water and abundant dissolved minerals. As pluvial lakes evaporate, they form brines, concentrated mixtures of a variety of dissolved salts and water. Great Salt Lake is one such briny pool, with salinity from three to ten times that of ocean water. During the final stages of desiccation in a pluvial lake, the most soluble minerals precipitate, including a suite of salts that gives the dry, flat surface its brilliant white color.

Playas exist throughout the Great Basin and come in a variety of sizes. Some playas hold unpalatable, salty water part or all of the year. Others hold no water and haven't for a very long time. Ongoing mining operations in many of these playas extract mineral deposits from the evaporites left behind by desiccating pluvial lakes. Economically valuable compounds within evaporites include sodium carbonate, sodium sulfate, phosphate, and borax, all used in cleaning products and many other types of products, too; lithium, used in medications and batteries; bromine, also used in pharmaceuticals, as well as in flameproofing and water purification; and halite, or table salt.

To produce thick evaporite deposits, an environment must meet four general conditions. First, the topographic and climatic conditions must be right for producing large lakes. Second, a source of saline material must exist within the contributing drainage basin. Typically, dissolved minerals come from weathered rock, but other processes also contributed within the Great Basin. Recent volcanism within the region brought an abundance of magma, or molten rock, to shallow depths below the surface. As molten rock solidifies, it releases magmatic fluids, very hot water full of dissolved minerals. The Great Basin is rife with faults, which act as conduits, carrying magmatic fluids to the surface in the form of mineral springs. These springs supplied dissolved minerals to the growing pluvial lakes, enhancing their potential to eventually produce valuable evaporite deposits. Third, a lake must be able to concentrate dissolved minerals. If a lake has an outlet or overflows, it sends both water and dissolved

Mud cracks in the central portion of a playa are larger and more consistent due to clay and evaporite content. Silt, present near the historic shoreline, is absent here. Keys for scale.

minerals into another basin, so its salinity won't increase. However, if a lake is in a closed basin, water leaves only through evaporation, a route that dissolved materials can't take. So the concentration of minerals in the water increases over time. Fourth and finally, climate must change in such a way as to either increase the rate of evaporation or decrease the amount of basinwide runoff so that the lake completely evaporates. As the water evaporates, increasing salinity causes minerals to be deposited in sequence, the order of which depends on temperature, pressure, and mineral concentrations. After the last ice age peaked, the climate at Lake Bonneville became warmer and drier, and the lake slowly evaporated over the next 14,000 years. The drying lake left behind a playa rich in valuable mineral deposits.

The series of low earthen berms at the southern edge of the playa aren't natural features; they're ponding dikes designed to trap saline water. Starting in 1985, some 3,000 acres of solar-powered evaporation ponds and an 8-mile collection conduit were used to mine halite and potassium sulfate from the Sevier Lake playa. Although the operation is presently inactive, there's still potential for gathering valuable minerals here.

In southern Utah's present-day hot and dry climate, evaporation rapidly sends most surface water back into the atmosphere. The two perennial remnants of Lake Bonneville in Utah, Great Salt Lake and Utah Lake, derive their water primarily from alpine runoff in the Wasatch Range. They are beautiful blue pools in an otherwise arid region. Imagine for a moment the cooler, wetter climate that existed here 15,000 years ago and a sea of blue filling the 20,000 square miles from Idaho to the Wah Wah Valley in southern Utah. That immense body of water was truly the Great Basin's greatest lake.

6 FOLDED ROCKS
The Virgin Anticline,
Quail Creek State Park

Much of southern Utah's spectacular scenery owes a great debt to the region's thick sequence of sedimentary rocks. Here at Quail Creek State Park, the layered sedimentary rocks of the Moenkopi and Chinle formations form the brightly colored cliffs that rise from the shores of the small reservoir.

Sedimentary rocks are, simply put, rocks composed of consolidated sediment. In other words, they're made up of particles produced by the

View from stop 1 northeast across the reservoir to the axis of the Virgin anticline

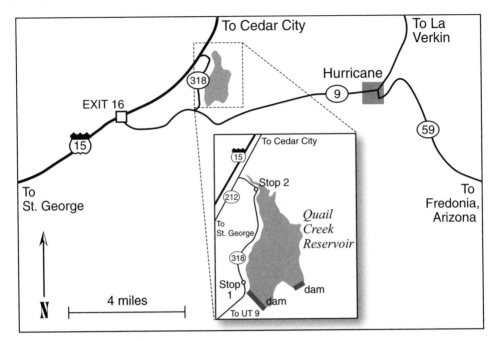

GETTING THERE

To reach Quail Creek State Park from the west, take I-15 north from St. George to exit 16 and proceed east on Utah 9. After 2.5 miles, turn left (north) onto Utah 318, the entrance road to Quail Creek State Park. Proceed 1 mile to a large gravel parking area on the right. This is stop 1, where we'll become familiar with the general geology of this area. Coming from the east, Utah 318 is 7 miles west of the intersection (in the town of Hurricane) of Utah 9 and Utah 59; turn right (north) on Utah 318 to reach the park. Stop 2 is a parking area on the right 1.6 miles farther north on Utah 318.

weathering and erosion of other rocks or soil. The particles may be as large as boulders or as tiny as molecules precipitated from water. As loose sediment is buried beneath other layers of sediment, processes such as compaction, cementation, and crystallization convert it to solid rock. These processes are known collectively as lithification. Most sediments are deposited in nearly horizontal layers, a concept referred to as the principle of original horizontality.

Across the street to the west from stop 1, there's a good view of the brightly colored layers of the Moenkopi formation and the overlying Chinle formation. Both formations date to Triassic time, having been deposited about 240 to 210 million years ago. Only the upper two members of the Moenkopi formation are exposed here. The red-and-white striped Shnabkaib member of the Moenkopi formation at the base of the cliff is the oldest unit in the sequence. This rock unit was deposited in a nearshore marine environment on a gently sloping continental shelf. Slight fluctuations in sea level caused the shoreline to migrate great distances and resulted in a complex depositional environment. The rocks formed from these nearshore sediments include siltstone, sandstone, gypsum, and thin beds of dolostone.

Carefully cross the street and climb a short distance up the low, gray hills at the base of the cliff to get a closer view of this unit and its varying rock types. Stay on the small trails and don't try this when it's wet. If you do, you may need a tow truck to pull you out of the muck and you'll probably have to leave your boots behind. These low hills consist predominantly of siltstone with interbedded gypsum, and if you keep your eyes on the ground, you may see small pieces of gypsum glistening in the sun. Gypsum is a mineral that can take various forms; the transparent type here is known as selenite. The gypsum and siltstone weather to a fine, powdery soil usually covered by a microbiotic crust, also known as cryptobiotic or cryptogamic crust. This crust is an assemblage of tiny organisms such as fungi, algae, cyanobacteria, mosses, and lichens that live on or just beneath the surface of the soil. They make the soil appear bumpy and darken its color as they mature. These delicate crusts are a very important part of arid ecosystems. They help

Rock sequence at stop 1 from lowest to highest: Moenkopi formation (banded Shnabkaib member, "Purgatory sandstone," and Upper Red member) and resistant Shinarump conglomerate of the Chinle formation

Thin veins of alabaster, a type of gypsum, in the Shnabkaib member of the Moenkopi formation at stop 1

hold the soil together, thus reducing wind and water erosion. Also, the rough surface they create catches and holds water, increasing infiltration and decreasing runoff. Once disturbed, they take many years to recover, so please try not to trample them.

Continue up to the base of the lowest red layer to see another form of gypsum called satin spar, which is fibrous. It occurs here in veins crosscutting the rock unit, and you may be able to see some pieces scattered around on the ground that have weathered out of the rock. This rock unit also contains a third type of gypsum, known as alabaster, which occurs as fine-grained white beds ranging in thickness from less than an inch to several feet. Thin beds of dolostone and sandstone are also present. These last two rock types are somewhat more resistant than the others and tend to form ledges throughout this rock unit.

Return to the parking lot and look back at the cliff. Above the Shnabkaib is the Upper Red member of the Moenkopi. It consists primarily of reddish brown to chocolate-colored siltstones and sandstones. This unit was deposited during a low phase in the changing sea level cycle, and the reddish color is caused by oxidation of iron contained in the rock. The relatively thick, yellowish band just above the Shnabkaib is the Purgatory sandstone, an informally named unit near the base of the Upper Red member in this area. You'll get a closer look at the Upper Red member at stop 2.

Sitting on top of this outcrop is the Shinarump conglomerate member of the Chinle formation. This rock unit consists primarily of sandstone and conglomerate and was deposited in a braided stream environment. It is very resistant to erosion and forms a cap on the cliffs here at Quail Creek.

The principle of original horizontality tells us that these sediments were deposited in more or less horizontal layers. Therefore, the rocks that formed from them should be horizontal strata. However, if you look at the cliff at the north end of the lake, you'll see that this is not always the case. The rocks there arch upward. Look toward the western end of the cliff, and you'll see that the layers of rock tilt down to the west. This inclination of strata is known as dip. Notice how the angle of dip increases farther west. Now look to the east. The rocks on that side tilt in the opposite direction. If you were to take this book, close it, and apply pressure to the edges, causing it to bow upward, it would look something like the structure in front of you. Sedimentary rocks may begin as horizontal layers, but as you see, things can change.

The earth's crust is a dynamic system. Forces such as compression and tension are constantly at work applying stress to crustal rocks. These stresses manifest themselves in the rocks as various geologic structures. Faults such as those described in vignettes 7 and 16 are breaks in crustal rocks and constitute one type of geologic structure. Bends in strata, known as folds, are another. The geologic structure you're looking at and indeed

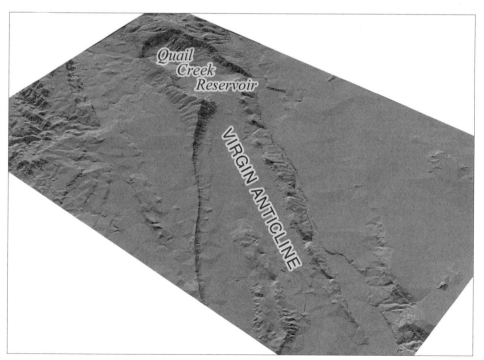

Three-dimensional image of the Virgin anticline

standing in the center of here at Quail Creek State Park is a type of fold known as an *anticline*, a word derived from Greek that means "opposite inclined." An anticline is a fold in which the convex part points upward. Anticlines usually result from compressive forces. In synclines, the concavity points upward, like a bowl, and the convex part points down. The structure before you in Quail Creek State Park, called the Virgin anticline (after the Virgin River, which passes nearby), is about 30 miles long, trends to the northeast, and was formed by compressive forces about 100 million years ago.

Look back toward the north. Imagine planing the top off the fold down to the level at which you are standing. The Shnabkaib member of the Moenkopi formation, the oldest unit exposed in the park, would occupy the center of the fold. From there, if you walked to either the east or the west, you would cross increasingly younger strata: first the Upper Red member of the Moenkopi, then the Shinarump conglomerate. Because the youngest rocks are the top layer and, in an anticline, the sides, or limbs, plunge downward, the oldest rock will always appear at the center of the fold. Synclines are the opposite. They have the youngest rock unit in the center of the fold. The parking lot and lake are actually located in the center of the anticline on the Shnabkaib.

A few terms are crucial in any discussion of folds. The hinge of a fold is the line of maximum curvature in the folded bed. Each individual bed in a fold has a hinge. The plane that connects all of the hinges is known as the axial plane. Look north again and see if you can locate where the beds in the fold curve most tightly. The surface connecting these points is the axial plane. Here, the axial plane is almost vertical. A fold in which the axial plane is nearly vertical is called a symmetrical fold. The sides of the fold are called the limbs. In an anticline, the limbs dip away from the axial plane.

Drive on to stop 2, where we'll discuss two other important aspects of folds. We've already mentioned one of these: dip, a measure of the angle between a rock layer and the horizontal plane. Look across the north end of the lake to the beds of Moenkopi and Shinarump. Take this book and hold it in front of you horizontally at eye level. Next, pick out an individual bed in the rock face. The angle that the bed makes with the book is the dip. The beds you're looking at dip at an angle of about 25 degrees. Another important concept is strike. The direction of a line formed by the intersection of a structure's surface and a horizontal plane is the strike. Strike is always expressed as a compass direction. Across the street, you can see an outcrop of siltstone and thin sandstone beds of the Upper Red member of the Moenkopi formation. Walk over, pick one layer, and hold this book against its underside in a horizontal position.

The line formed by the intersection of the edge of the book and the rock layer shows the direction of strike. In this case, the line trends more or less through you and across the street in a northeast direction. While we're here we can also get the dip. Dip is always measured in a vertical plane at a right angle to the strike. The dip, then, would be the angle between the book and the bed measured at right angles to the direction of strike. The strike and dip combine to define the three-dimensional orientation of the bed, known as its attitude.

The sediments that formed these rocks were deposited horizontally, but now the rocks are folded in an anticline. To understand what happened here in the intervening millennia, we have to step back for a view of the bigger picture.

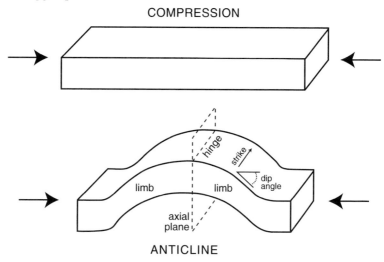

Compression and anticline formation

About 400 miles due west of where you stand, the North American Plate meets the Pacific Plate along the California coast. When two plates come into contact with one another, one of three general scenarios plays out. The plates can smash into each other head-on, piling up the earth's crust into huge mountain ranges such as the Himalayas. One plate can slide under the other; this is called subduction. Or, in what's known as a transform boundary, the two plates slide alongside each other. Southern California's San Andreas fault is a transform boundary.

About 100 million years ago, during Cretaceous time, the western edge of the North American Plate was west of here somewhere in present-day Nevada and an oceanic plate, the Kula Plate, was sliding under it. One effect of this subduction was the formation of a string of volcanoes just inland from the subduction zone. A modern-day example of the same system can be found along the western coast of South America, where the oceanic Pacific Plate is being subducted beneath the continental South American Plate. The volcanoes of the Andes are the surface expression of this subduction.

As the North American Plate overrode the Kula Plate, the subduction also caused intense compression of crustal rocks to the east. These rocks folded, and large sheets of rock broke off and slowly thrust eastward. It was this folding and thrusting that caused the formation of the Virgin anticline and many other features in the area. This particular episode of folding and thrusting is known as the Sevier Orogeny. Orogeny is the term for the folding, faulting, and thrusting that forms mountain ranges. As the mountains built upward, they shed sediments that large rivers transported northeast and deposited in an inland sea, resulting in the thick sequence of Cretaceous shales and sandstones found in southern Utah today. The Virgin anticline, laid out before you here at Quail Creek State Park, is a beautiful expression of compression during the Sevier Orogeny so long ago.

Looking south along the west limb of the fold south of Utah 9

7 IN THE HEART OF THE TRANSITION ZONE
Hurricane Fault

Every year countless travelers leave I-15 and proceed east on Utah 9 to Zion National Park. Most motor on through without giving their surroundings more than an occasional glance, intent on reaching their destination. This is unfortunate, as there are a number of interesting and easily accessible geological features along this route, including the Virgin anticline (see vignette 6), the Harrisburg Dome, and numerous

View northwest from the overlook. The tilted beds in the center, of Triassic age, have been dropped down along the Hurricane fault. The blocky rocks to the right are older and belong to the Permian Kaibab formation. On the far right, the Triassic Moenkopi sits on top of the Kaibab.

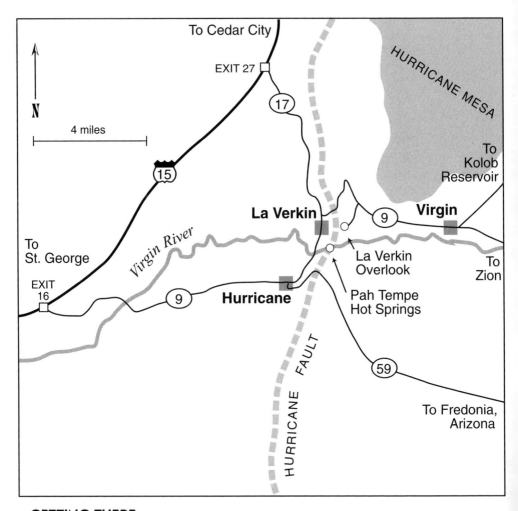

GETTING THERE

To get to the La Verkin Overlook from the west, if you're traveling north on I-15, exit at Utah 9 (exit 16) and proceed east to the town of Hurricane. Continue on Utah 9 through La Verkin to the intersection with Utah 17. Stay on Utah 9, which goes to the right and climbs the Hurricane Cliffs, and travel 2 miles east of the intersection to the turnoff to the overlook, which is a signed dirt road on your right. The overlook is at the end of the road, 1.3 miles from the turnoff. Traveling south on I-15, take Utah 17 (exit 27) 6 miles south to the junction with Utah 9. Turn left onto Utah 9 and follow the directions above to the overlook. From the east, take Utah 9 out of Zion National Park. The turnoff to the La Verkin Overlook will be on your left 17 miles west of the park entrance.

relatively recent basalt flows. Just east of the town of La Verkin lies another feature, the Hurricane Cliffs. Stretching for nearly 50 miles and reaching up to 1,000 feet in height, the cliffs are actually the expression of the Hurricane fault. The cliffs are also a message, if you know how to read it: this area is prone to earthquakes.

Follow the directions in "Getting There" to the La Verkin Overlook, walk to the edge, and look west. Some faults are cryptic and are discovered only through detailed fieldwork and mapping. Others need little or no introduction. Judging from the view before you, the Hurricane fault surely falls into the latter category. But how do we know for certain that this feature is actually a fault scarp? Is there anything visible here that could help us determine this?

The light-colored rock here at the overlook is Permian limestone of the Kaibab formation, and it is this resistant rock unit that forms the imposing Hurricane Cliffs below. Look to the northeast to Hurricane Mesa in the distance. The red rocks exposed along the edge of the mesa belong to the Moenkopi and Chinle formations from Triassic time. That seems about right: Triassic rocks sitting on top of older rocks of Permian age. Now take in the view to the west. There you will see the same rock units, but over 1,000 feet lower, exposed along the Virgin anticline. The fact that you're standing on Permian limestone with younger rock below you is all the information you need to infer that a fault is present here.

Hurricane Mesa displays a complete section of the Moenkopi formation.
The Shinarump conglomerate member of the Chinle formation caps the mesa.

Further exploration along the scarp in this area would produce more evidence that faulting has produced this feature. One bit of evidence is the exquisite slickensides present in the Kaibab formation along Utah 9 and in some small drainages south of Hurricane; another is the angled beds of the Moenkopi and Chinle formations along the base of the scarp near Pah Tempe Hot Springs, dipping steeply to the west as a result of drag along the fault.

The Hurricane fault is the largest of the three faults in the transition zone between the stable Colorado Plateau and the heavily fractured Basin and Range geomorphic provinces. (The other two faults are the Sevier and Paunsaugunt faults.) Total offset along the Hurricane fault generally increases from south to north along its 160-mile length. Here near the town of Hurricane, the total offset is estimated to be nearly 8,000 feet.

Movement along the Hurricane fault is estimated to have begun in early Pliocene time, about 5 million years ago. If you look down toward the town of La Verkin, you'll see some areas of black basalt. Numerous relatively recent basalt flows have occurred in this region, and some of them have been offset by the fault. Several studies of these basalts and of surface ruptures in the area indicate that the latest significant offset

In this drainage south of Hurricane, Triassic rocks on the left have dropped down and are in contact with the older Kaibab limestone, of Permian age. Darker streaks at the bottom right are pulverized remnants of the Triassic rocks plastered onto the limestone face, a type of material called fault gouge.

occurred fairly recently, probably in late Pleistocene time. This leads to the conclusion that the Hurricane fault is still active in an area where residential and commercial developments are being built at a rapid, even alarming, rate. Obviously, an active fault of this size in a burgeoning population center may lead to some problems down the road. Future earthquakes are probable, although how big they'll be and when they'll occur are questions that cannot as yet be answered. But what exactly are earthquakes, and why do they occur?

An earthquake is a response to stresses that build up within the earth's crust. These stresses tend to build in areas where crustal plates interact with each other. Here at the western edge of the Colorado Plateau, the crust is slowly pulling apart, and this crustal extension becomes more intense as you head west into the Basin and Range. As the stress slowly builds over time, the rocks begin to deform, or bend. This deformation can take one of two forms, depending on whether it occurs near the surface or down deep. Rock near the surface tends to be brittle. If only a slight amount of stress is applied to it, the rock will deform; and if this minor stress is removed, the rock will return to its original shape. This type of behavior is called elastic deformation. If the applied stress is too great, the rock will break. Rock at greater depths behaves differently. When stress is applied to these rocks, they too begin to deform, but if the stress is removed, they don't return to their original shape. This is termed plastic deformation.

Generally, as stress slowly builds, the brittle rocks at the surface begin to deform. The more the rocks deform, the more energy they store. This straining can't go on indefinitely, and sooner or later the rocks fracture and slip past one another. When the rocks fracture, some of the stored energy is released, partly as heat and partly as waves that transmit energy away from the rupture. These waves are what we feel as an earthquake. The physical point at which the rupture begins is called the earthquake's focus. The epicenter is the point on the earth's surface directly above the focus. The plane along which the rocks slip past each other is the fault.

Different types of forces within the earth's crust create different types of faults. Tensional forces pull the crust apart, causing the block above the fault, known as the hanging wall, to slide down relative to the block below the fault, known as the footwall. This is called a normal fault. Compressional forces, on the other hand, push the hanging wall up relative to the footwall, resulting in what is known as a reverse fault. A third type of force, shear, causes the blocks to slide past each other horizontally, resulting in strike-slip faults.

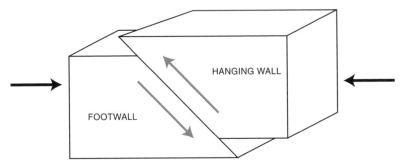

Compression results in a reverse fault.

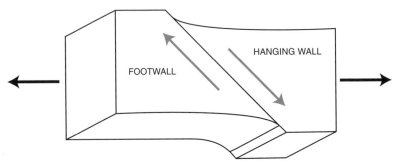

Drag along the fault bends beds in a normal fault.

As mentioned earlier, the crust in the transition zone between the Colorado Plateau and the Basin and Range is being pulled apart. The Hurricane Cliffs are a direct result of this stretching. As the crust is pulled apart, the area to the west of you, the western fault block, is dropping down relative to the block upon which you are standing. This cliff is actually a large normal fault that has formed as a result of thousands of separate movements over the past 5 million years. With each movement, the western block drops down a little more and the cliff gets higher.

Historical records of seismic activity in this area include twenty or so earthquakes with a magnitude greater than 4.0 on the Richter scale during the last century. The largest of these were a magnitude 6.3 quake near the Pine Valley Mountains in 1902 and a magnitude 5.8 quake on September 2, 1992, southeast of St. George. These are not large-magnitude quakes, but they do suggest a significant seismic hazard for the area. Seismic hazards associated with earthquakes can take a number of forms, including damage due to ground shaking, slope failure, and rockfalls. Obviously these hazards increase with the size of the earthquake.

Seismologists determine earthquake size in two ways. One method measures the intensity, or amount of shaking, caused by the earthquake. In this method, intensity is determined by the degree of damage to structures, the amount of disturbance to the surface of the earth, and the extent of animal reactions to the shaking. Before instruments capable of quantitatively measuring magnitude were developed, this was the only method available to compare the sizes of various earthquakes. Earthquake intensity is generally described using a twelve-degree scale based on the one developed by Guiseppe Mercalli in 1902. The modified Mercalli scale, as it is called, ranges from I, not felt by people, to XII, total destruction. One drawback of this scale is that intensity ratings tend to be rather subjective and depend on factors such as population density and the type of building construction in the area. For example, an earthquake in a heavily populated area might receive a higher intensity rating than an earthquake of similar Richter magnitude in an unpopulated area. Insurance companies use the Mercalli scale to arrive at estimates of potential earthquake damage.

When instruments capable of quantitatively measuring the energy released by an earthquake became available, seismologists developed a standard scale to compare earthquakes worldwide. This scale, originally introduced in Japan, was later modified by Charles Richter in 1935 and has become the industry standard that now bears his name. It uses wave heights, or amplitudes, as measured by a seismograph to calculate the magnitude of the earthquake. Because the sizes of earthquakes vary enormously, the wave amplitudes can differ by factors of thousands from one event to another. To compensate for this, Richter used a logarithmic scale in which each unit of magnitude corresponds to a tenfold increase in amplitude. This means that an earthquake with a magnitude of 6.0 is ten times more powerful than one with a magnitude of 5.0. The Richter scale of magnitude has no fixed upper limit, although crustal rocks certainly do have limits to their strength. Even the strongest rocks can only store a given amount of energy before they fracture.

On September 2, 1992, the people of southwestern Utah experienced firsthand the hazards of living in a seismically active area when a magnitude 5.8 earthquake shook the region. (Modified Mercalli intensities in the range of VI to VII were assigned to this earthquake.) The epicenter was about 12 miles southeast of St. George, and the focus was at an approximate depth of 10 miles. It is thought that this earthquake occurred along the fault plane of the Hurricane fault. Generally, ground shaking is the most damaging and widespread hazard associated with earthquakes, but during the St. George event, relatively little damage occurred in this way. A number of older, unreinforced masonry buildings showed minor structural damage in the area of maximum intensity, but cracked chimneys and fallen plaster were the extent of this type of damage in St. George. No surface ruptures have been located in association with this earthquake, but the earthquake did trigger a large slope failure in the town of Springdale, about 25 miles east of the epicenter (see vignette 10). Numerous rockfalls were also reported along the steep cliffs in the area, including the one you're standing on. Look below you at the buildings near the base of the cliffs. These structures and any people in them are in imminent danger from future rockfalls.

So how can we determine the earthquake potential of active faults? A primary method is to study the patterns of a fault's past behavior. It isn't enough to simply look at earthquakes in the past couple of centuries. Many of the most dangerous fault segments may not have ruptured during that short time period. It's a simple fact that strain energy increases with increasing time between earthquakes, and this leads us to the frightening conclusion that the quiet sections of active faults may in fact pose the most danger. So we must look further back. To do this, geologists dig trenches across a fault and map the sediments and soils. They then measure the displacement of the various layers across the fault. Using radiocarbon dating, they also determine the youngest sediments that have been displaced. This study of the history of activity along faults, referred to as paleoseismology, provides important information about the frequency and magnitude of earthquakes along different segments of active faults.

Maps have been developed to show the seismic hazards across the United States. The Uniform Building Code, an attempt to standardize all aspects of building construction, includes a map of seismic zones based on these seismic hazard maps. Southwest Utah falls within zone 2B, and all buildings constructed in this area must comply with the regulations stated in the Uniform Building Code for that zone. Southern California lies within zone 4, the highest level of hazard, and buildings constructed there must conform to more stringent regulations. California has the most rigid building code in the nation, developed from studies of recent earthquakes such as the 1971 San Fernando earthquake. Many relatively modern structures were extensively damaged during this earthquake, despite having been built to code. Because of this, the Uniform Building Code was extensively modified shortly thereafter. Future modifications may well be required after completion of studies of structural damage during other recent earthquakes: the 1989 Loma Prieta earthquake, the 1994 Northridge earthquake, and the 1995 Kobe, Japan, earthquake. Older structures built before modern code modifications tend to be the most hazardous.

Often it's only after an earthquake that government agencies and the public start thinking about earthquake safety. After all, it takes money to conduct preventive studies and reinforce existing structures that are at risk. But this is shortsighted, especially when you consider the billions of dollars in damages caused by the Northridge earthquake alone—an earthquake classified as moderate. Be proactive. Find out what seismic zone you live in, and if earthquakes are likely, take steps to be prepared. The fact that no earthquakes have happened recently is no guarantee that one won't happen tomorrow.

Seismic hazard map for the western United States

Slickensides on Kaibab limestone south of Hurricane

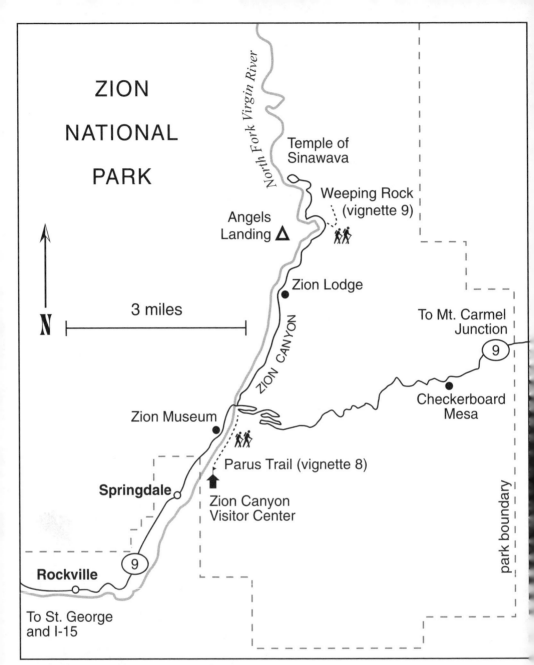

GETTING THERE

From St. George, take I-15 north 7 miles to exit 16. Proceed east on Utah 9 for 35 miles to the south entrance of Zion National Park at the eastern edge of the town of Springdale. Once inside the park, take the first right and proceed to the visitor center parking lot. From the east on U.S. 89, go 42 miles south from Panguitch or 17 miles north from Kanab to Utah 9 and proceed west about 16 miles to the east park entrance. Once inside the park, continue to the turnoff for the visitor center. The Parus Trail starts from the parking lot and is about 3 miles round-trip. It's the only trail in the park that allows bicycles and dogs (but only on leashes).

8 AN ANCIENT DESERT SLEEPS
The Navajo Sandstone,
Zion National Park

Zion National Park is home to many rock units with varying physical characteristics, but one particular unit is chiefly responsible for the park's unique and stunning beauty: the Jurassic Navajo sandstone. Nowhere in the park are you out of sight of magnificent, tall cliffs of this colorful sandstone. Zion Canyon reflects the convergence of sturdy rock, regional uplift, and the erosive power of the Virgin River, a southward-flowing tributary of the Colorado River that, since completion of Hoover Dam in 1935, now spills into Lake Mead. Because the Navajo sandstone is the most visible rock unit in the park, and the most important in the canyon's development, there are many fine places to make its acquaintance. One is the Parus Trail, which follows the Virgin River upstream from the visitor center, crossing the river several times before reaching the turnoff for the Zion Canyon Scenic Drive. Although the Virgin River seems unassuming (except at full flood stage), it has carved thousands of feet down through strong rock to create the canyon spread before you. Time and moving water are powerful allies in shaping the earth's surface. If you'd like to see the Navajo sandstone up close, and you welcome a physical challenge and don't fear heights, hike the trail up to Angels Landing. It zigzags up a sheer face of the sandstone to a rock pedestal with fantastic views up and down the valley.

Stroll north along the Parus Trail past South Campground (oddly enough, the northernmost campground in the park). The towering cliffs of Navajo sandstone that border Zion's central valley, in which this trail lies, consist of over 2,000 feet of rock grading in color from dark brown at the bottom to pink then white near the top. The color comes from

The entire vertical sequence of Navajo sandstone is on display in virtually every direction.

iron oxides that, along with calcium carbonate, cement individual sand grains together. The Navajo sandstone sits atop the Kayenta formation, visible in the slopes at the base of the valley floor; these slopes are partly covered by rockfall debris from above. Like the lower portion of the Navajo sandstone, the Kayenta formation is dark brown, but in contrast to the Navajo sandstone, it consists of bedded siltstone and sandstone, so it's less internally consistent. Its inconsistent structure causes the Kayenta formation to weather and erode more rapidly than the Navajo sandstone, creating more gradual slopes than those above.

The Navajo sandstone is a remnant of a huge arid region called the Navajo Desert, a sand sea, or erg, that existed about 200 million years ago during Jurassic time and was larger than today's Sahara. Sandstone remnants of this great desert can be found from southern Nevada to Wyoming. To understand the Navajo sandstone, we should first talk about the relationship between climate and deserts. Why, for example, are there abundant deserts in the southwestern United States but humid climates in the southeast? Why is the world's driest desert, South America's Atacama Desert, next to an ocean?

The answers lie in the fundamental relationship between the earth and the sun and the geometry of continents and oceans on the earth's surface. It all begins at the equator. The planet's equatorial region is closer to the sun than the poles, which means the equator receives more solar energy. This abundant solar energy does two things: it heats the atmosphere, and it causes high rates of evaporation, creating very humid air. This hot, humid air rises in a uniform band called the equatorial low, so named because rising air creates low surface atmospheric pressure. But rising air cools rapidly. Cooling a humid air mass is analogous to squeezing a moist sponge: water is lost via precipitation, and the air (the sponge) dries out. In the equatorial rain forests, it rains every day due to the "wringing out" of constantly rising air.

This now dry air is pushed north and south from the equator by the hot, humid air that continues to rise there. By the time the dry air reaches about latitude 30° north and latitude 30° south, it has cooled again and descends. This very dry air warms as it falls, creating two bands of high pressure called the subtropical highs. The subtropical highs

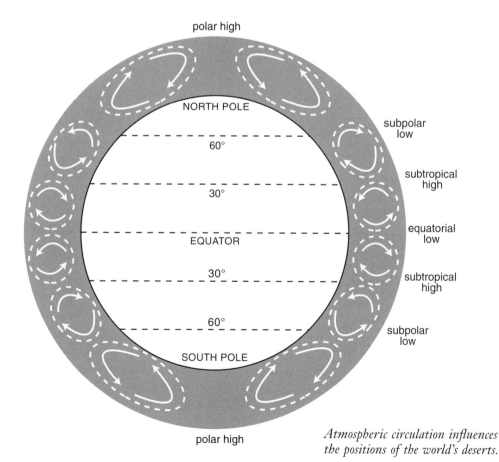

Atmospheric circulation influences the positions of the world's deserts.

are the leading cause of arid regions; a quick look at a world atlas will confirm that most modern deserts lie at or very near latitude 30° north or 30° south. The Mojave, Sonoran, and Chihuahuan Deserts in North America and the Sahara and Arabian Deserts lie at latitude 30° north. The Atacama and Kalahari Deserts and Australia, which is 85 percent arid, lie at latitude 30° south.

There's another way to cool an air mass and wring out water. As air travels over mountains, it cools. Since cool air can't hold as much water as warmer air, it loses water through precipitation. This dynamic is also a major contributor to aridity in the deserts of the southwestern United States. In this region, air masses generally travel from southwest to northeast (the dominant wind direction between latitudes 30° and 60° north), passing over many high mountain ranges along the way. The windward side of each range receives more precipitation than the lee side, and each range that the air mass travels over receives progressively less moisture. Regions that receive little precipitation due to interception of moisture by mountains are said to lie in the rain shadow of those mountains, and this is certainly the case for the Great Basin and the Colorado Plateau.

But a question arises: Florida lies at latitude 30° north, but anybody who's spent a muggy summer day there knows that it isn't dry. What's going on? Ocean currents play a role in creating arid climates as well. The earth's rotation, counterclockwise when viewed from the north pole, imparts motion to the world's oceans. Large-scale currents in northern hemisphere oceans flow clockwise, while currents in the southern hemisphere flow counterclockwise. Ice-cold arctic water is carried southward along the west coast of the United States, and warm Caribbean water is carried northward along the east coast. When wind blows from a warm ocean surface onto land in the eastern United States, the air cools and its relative humidity rises. When wind blows from a cold ocean surface onto land in the western United States, the air warms and its relative humidity drops. Warming, drying air never produces rain. Thus, the southwestern United States is predominantly arid, while the southeastern United States is humid.

About 200 million years ago, the Navajo Desert sat at latitude 30° north on the west coast of a northern hemisphere continent, much as the Sahara Desert does today. The central Sahara is full of blowing sand and dunes, and so was the Navajo Desert. We know this by looking at sedimentary structures called crossbeds in the Navajo sandstone. Dunes, created by abundant blowing sand, have a gently sloping windward face and a steeper downwind, or lee, face. Dunes move downwind as sand grains blow up the windward face, then slide down the lee face. (We discuss modern dunes in more detail in vignette 11.) Crossbeds are layers that form as sand slides down the lee side; if you were to cut a modern dune in half vertically, you would see many sloping layers all angled down toward the dune's lee face. The dip direction of crossbeds essentially tells us which way the wind was blowing when these grains were deposited (for more about dip, see vignette 6). Crossbeds in the Navajo sandstone are very large and could only have formed in a sand sea with dunes as high as 100 feet.

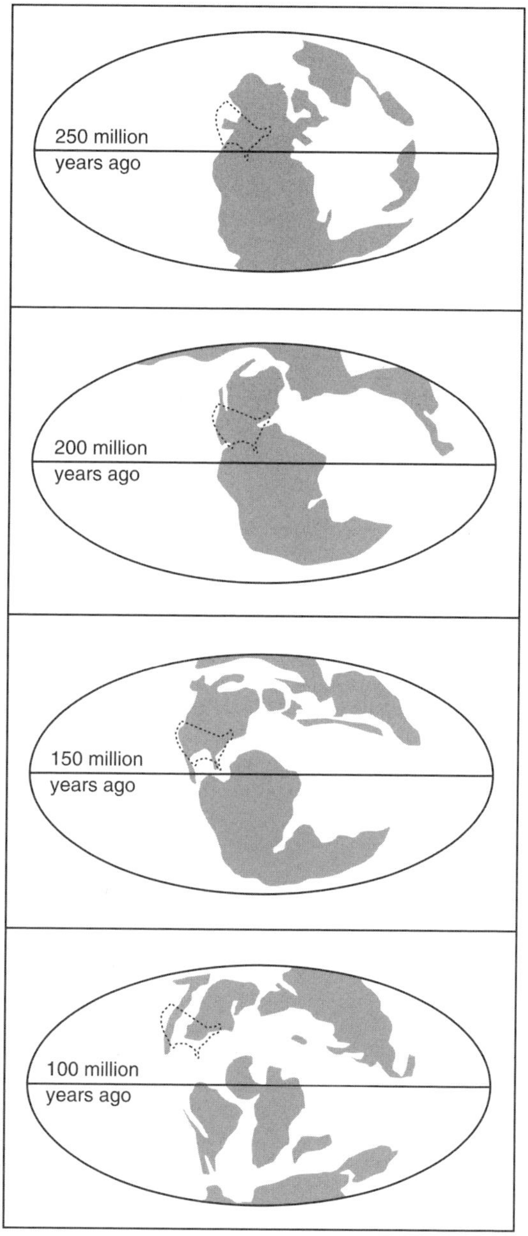

Late Paleozoic and Mesozoic positions of continental landmasses. Present-day southern Utah sat at latitude 30° north—often a zone of intense aridity.

250 million years ago

200 million years ago

150 million years ago

100 million years ago

Periodically, wind action removed sand in parts of the Navajo Desert and deposited it elsewhere, interrupting dune formation in some places. These interruptions separated one group of crossbeds—strata consistent in sand movement and deposition—from another. In the Navajo sandstone, these groupings are typically 10 to 20 feet thick. As you approach the northern end of the Parus Trail, look at the Navajo sandstone on either side of the trail and see how the dip direction varies between crossbeds, reflecting changes in ancient wind directions and velocities. You may also see thin beds of whitish limestone, the remains of basins between dunes that filled with water and subsequently dried up. It is on the margins of these small lake deposits that the occasional dinosaur footprint appears in the Navajo sandstone.

Many changes have taken place on our planet since the time of the Navajo Desert, and some of them are recorded within the Navajo sandstone. Southern Utah was flooded by the sea, dried out, then flooded again. This is probably when lithification—in this instance, the transformation

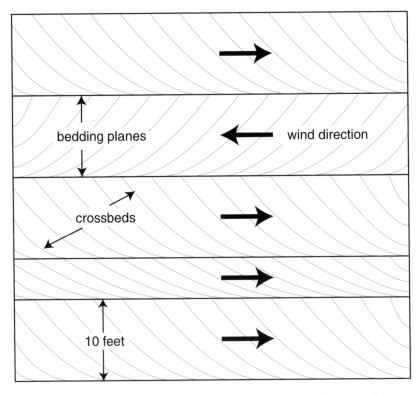

Crossbeds in Navajo sandstone indicate wind direction at the time of dune formation. The strata here are approximately 10 to 20 feet thick.

of sand into rock—began, as water percolating from above cemented the unit and stained it with iron oxide. Later, horizontal stresses associated with large-scale plate tectonic motion raised the land surface and left their mark as large-scale fracture systems visible in aerial photographs. We know this occurred after lithification because soft sediment flows in response to stress, whereas hard rock breaks. The fractures tell us the Navajo sandstone was already rock when the stress occurred.

Still later, the sandstone rose as part of the uplift of the Colorado Plateau (see vignette 24 for more on this uplift), and erosion began to remove overlying rock. Removal of overburden decreased pressure on the underlying rock and allowed it to expand, creating new, smaller

Aligned fractures indicate that regional tectonic stresses occurred after lithification.

fractures like those visible on Checkerboard Mesa, near the park's east entrance. Erosion by the Virgin River then carved out Zion Canyon, leaving today's towering cliffs of sturdy Navajo sandstone.

The former sand sea here is long gone, and the topography has changed substantially. Where an ancient ocean once stood to the west, the continent has grown. Mountain ranges that have arisen to the west now block the flow of moisture inland, and the desert displays fractured bedrock surfaces and deep canyons instead of vast expanses of blowing sand. But the modern climate here is still strongly influenced by the subtropical high and by cold ocean water to the west, just as it was in Jurassic time. It's still hot and dry in this twenty-first-century desert superimposed on the skeleton of its ancient kin.

Checkerboard Mesa displays a pattern of horizontal crossbeds and vertical fractures. The latter were probably caused by expansion as erosion removed the immense weight of overlying rock.

Crossbeds in Navajo sandstone reveal the direction of winds in the Mesozoic Navajo Desert.

The view south from Angels Landing reveals the Navajo sandstone in all its glory.

GETTING THERE

From St. George, take I-15 north 7 miles to exit 16. Proceed east on Utah 9 for 35 miles to the south entrance of Zion National Park (see map, p. 58) at the eastern edge of the town of Springdale. Once inside the park, take the first right and proceed to the visitor center parking lot. From the east on U.S. 89, go 42 miles south from Panguitch or 17 miles north from Kanab to the junction with Utah 9. Proceed west on Utah 9 about 16 miles to the east park entrance. Once inside the park, continue to the turnoff for the visitor center. From April through October, passenger cars are prohibited on Zion Canyon Scenic Drive, but a shuttle bus takes you from the visitor center parking lot to sites in the park. Be warned that during summer the parking lot may fill by 10:00 a.m., so get there early. Shuttle buses leave every six to eight minutes. Get off at the Weeping Rock stop. From November through March, you can drive up Zion Canyon Scenic Drive. In this case, make the turn up Zion Canyon and proceed north past Zion Lodge to the Weeping Rock parking area. The trail to Weeping Rock is short and paved but rather steep. Binoculars will be useful for part of our discussion, so bring them if you have them.

The Weeping Rock alcove seen from the parking lot

9 WATER FROM STONE
Weeping Rock, Zion National Park

Water—we tend to take it for granted. Just turn on the faucet and there it is. But spend some time out in the arid country of southern Utah and you might start giving this cool, precious fluid a little more thought. Take away the convenience stores that have popped up in every little town, their refrigerators filled with bottles of the stuff, and you might give it a lot of thought. Take a walk along a steep, rocky trail without taking enough of it along, and suddenly water may become the most important thing in your life.

At first glance, you might think that much of the landscape of southern Utah is nearly devoid of water, but this is not really the case. Water is actually present in most desert regions; the problem usually is finding and extracting it. The most common interface between these hidden water sources and the thirsty hiker is the spring or seep. Weeping Rock in Zion National Park is a beautiful example of a seep.

As you begin walking up the trail to Weeping Rock, notice the abundance of green that surrounds you. Zion National Park has a rich diversity of plant life, with approximately nine hundred different species within its boundaries. The Park Service has labeled many of the trees and shrubs along the trail, including bigtooth maple, Gambel's oak, and box elder, so this short walk is a good opportunity to get familiar with some of the park's plant life. As you continue up the trail, you may notice a slight rise in humidity. You'll soon find out why. Near the end of the trail, look up. A wet, dripping, glistening wall of rock covered with green rises before you—a scene that seems to have more in common with the tropics than with southern Utah. Why all this water? Why here?

69

A constant stream of water drips from the alcove's lip. The water acts as an evaporative cooler on a hot summer day.

When precipitation falls on the high country above this canyon, much of it quickly runs off the surface. Traveling through a system of drainages, it ends up flowing down the Virgin River. If you ever visit this park on a rainy day, don't just hide in the visitor center watching slide shows. Instead, put on a raincoat and get out there and watch as water gushes over the towering cliffs in spectacular fashion on its way to the river.

While most of it runs off, some of this precipitation seeps down through surface soils and sediment and into the rock beneath, where it is stored in small spaces between the rock grains. Rocks with many spaces between the grains are said to be porous. Porosity is the measure of the amount of these spaces in relation to the total volume of the rock, and it is usually expressed as a percentage. Many igneous rocks, such as granite, have porosities as low as 1 to 3 percent, and their ability to hold water is very poor. Other rocks—sandstones, for example—have porosities exceeding 30 percent. These rocks hold large quantities of water.

Porosity alone, however, will not get the water from the highlands above to this seep at Weeping Rock. In order for that to occur, the spaces between the grains must be connected, allowing water to flow slowly through the rock. Rock or sediment that allows fluid to move through it is said to be permeable. Some rocks may have high porosity, but because their pores aren't interconnected, water will not move through them, so their permeability is low. Pumice, formed during explosive volcanic eruptions, is an example of this type of rock. It's so full of air spaces that sometimes it floats. But because these air spaces aren't interconnected, water can't move through the rock. Fractures or joints in a rock can also influence its permeability by allowing fluids to move along the cracks, making the rock more permeable. This is why highly fractured granite may actually have relatively high permeability even though its porosity is low.

Like most of the rock towering above you in Zion Canyon, the wall in front of you is composed of Navajo sandstone. The quartz grains that make up this sandstone were once part of an immense dune field that stretched across most of southern Utah. Over time, subsequent deposits buried and compacted these sand dunes, and the sand cemented together to form sandstone (see vignette 8 for more about this story). Sandstones tend to have high porosity and good permeability, so fluids move through them with relative ease. The term effective porosity refers to interconnected spaces within rock; the average effective porosity of the Navajo sandstone exceeds 20 percent, a fairly high value. This rock unit is also highly fractured, thus increasing its permeability. Rock units that store and transmit large quantities of water are called aquifers. The Navajo sandstone is the main aquifer supplying much of the drinking water in southwestern Utah.

The upper section of the Navajo sandstone is almost exclusively quartz sandstone. Water that infiltrates at the top percolates down through the unit with the help of gravity. Toward the base of the unit, closer to the contact with the underlying Kayenta formation, thin layers of siltstone and shale begin to appear. These types of rock are much less permeable than the sandstone and form a barrier that the water can't penetrate. The water then flows downslope along the surface of this layer until it emerges as a spring or seep like the one here at Weeping Rock.

Walk to the north end of the observation platform. At about eye level in the rock wall, you'll see a thin layer of reddish rock. This is one of the relatively impermeable siltstone beds. If you look closely, you can see water emerging along the upper surface of this bed.

When water emerges along the face of a cliff, as it does here, it tends to weaken the rock, causing it to erode more easily. Over time, the rock begins to crumble, eventually collapsing to form an overhang, or alcove, and furnishing a unique microhabitat for plants and animals.

This structure, known as a hanging garden, provides a constant source of water that is often protected from the harsh desert sun. Many of the plants that grow in these hanging gardens are found nowhere else in the world. So observe them, photograph them, and enjoy them, but please do not disturb them.

The hanging garden environment before you can be divided into three growing areas, and specific plants are adapted to each area. We'll start at the top, at the seep layer, where the water first emerges from the rock. If you stand at the bottom of the small staircase near the pool of water and look up, you can see this area right at the uppermost border of the wet rock. In this zone, the water source is constant but there's very little soil. Algae is usually abundant, along with a few other plants that can root in the minimal soil or directly into the algal mat. Calcium carbonate, a mineral deposit most of us know as the white crud on our pipes and faucets, is abundant here at Weeping Rock, forming lumpy white deposits that help plants take root on the rock wall. If you look carefully (binoculars help), you may see a few maidenhair ferns and some scarlet monkeyflowers hanging on up there.

Directly below the seep layer, there's usually a vertical wall. Look up and to the left for a good view of this second type of growing area. A few varieties of brownish and greenish algae dominate this area, where water is still plentiful but soil is practically nonexistent. With careful observation, you can see little tufts of monkeyflower and columbine taking root in small cracks or in areas where calcium carbonate has precipitated.

The third growing area occurs where the vertical wall changes to a slope. This is by far the most diverse of the three areas, and here at Weeping Rock the observation platform has been built into this zone. Water is still abundant, and there's plenty of soil for plants to take root in. Back up on the observation platform, begin at the south end and walk slowly north, looking along the small ledges in the rock wall. This way you can observe, close up, some of the plants already mentioned and a few new ones. At the south end of the platform, alcove columbine is quite plentiful. This is one of the largest flowering plants found in this habitat. Its leaves are lobed, and the large white to pale yellow flowers bloom throughout the summer. Another plant common here, scarlet monkeyflower, can be seen a little higher up on the wall at this end. The leaves of this plant are oval with small teeth along the edges. The bright red flowers are quite conspicuous and bloom from late spring until midsummer.

From late April to earliest June, you'll find beautiful Zion shooting stars blooming on the small ledges of this part of the alcove. This flower is purple with a white and yellow center. Even when it's not in bloom, you can identify it by its rosette of broad, spatula-shaped leaves. Other plants commonly found here include maidenhair fern and the delicate white Zion daisy.

Toward the north end of the platform, large deposits of calcium carbonate have accreted on the rock face. This mineral is dissolved in the springwater, and when water molecules evaporate at the surface, the mineral is left behind. Eventually, molecule by molecule, grain by grain, it builds up to form deposits. These deposits are important in the first two growing areas described, giving the plants something to root in where not much else exists.

After many years of wandering the open spaces of the American West, we've experienced few pleasures as rewarding as stumbling upon a spring or seep while on a long hike up a hot, dusty trail. The opportunity to exchange the warm, slightly plastic-flavored water in one's backpack for the cool, sweet fluid emerging from the rock is always a welcome joy (though even springwater should generally be treated, as it takes some experience to recognize water that is safe to drink). As you observe the plants and animals living in this alcove under its curtain of dripping water, consider the many geologic processes that have occurred over millions of years to make this unique scene possible.

Bob Wieder investigates monkeyflowers and calcium carbonate deposits at Weeping Rock.

Maidenhair fern in calcium carbonate deposit

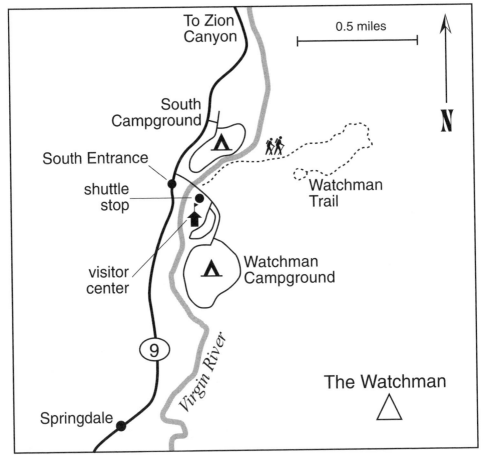

GETTING THERE

The viewpoint for the Springdale landslide is located at the end of the Watchman Trail in Zion National Park (see map, p. 58). From St. George, take I-15 north 7 miles to exit 16. Proceed east on Utah 9 for 35 miles to the south entrance of Zion National Park, at the eastern edge of the town of Springdale. Once inside the park, take the first right and proceed to the visitor center parking lot. From the east on U.S. 89, go 42 miles south from Panguitch or 17 miles north from Kanab to the junction with Utah 9. Proceed west on Utah 9 approximately 16 miles to the east park entrance. Once inside the park, continue to the turnoff for the visitor center. Be warned that during summer the parking lot is often full between 10:00 a.m. and 3:00 p.m., so don't plan to arrive during those hours. From the parking lot, walk to the Watchman trailhead, located at the bridge across the Virgin River just northeast of the shuttle bus stop. The trail, about 2 miles round-trip with less than 400 feet of elevation gain, is mostly easy with a few steep parts. During summer, the walk is coolest during the early morning or later evening. Binoculars are useful for part of our discussion, so bring them if you have them.

10 GRAVITY AT WORK
The Springdale Landslide

Gravity doesn't sleep. Gravity is one of the few things in life you can always count on. At the visitor center in Zion National Park, you stand surrounded by cliffs of sandstone that tower 2,000 feet above you. Gravity will not allow this scene to persist forever, though. Rest assured that one day these walls will be broken down and carried away, and gravity will have had a hand in it.

Anyone who has taken the shuttle-bus ride up Zion Canyon has heard the story of the landslide that occurred there in April 1995. The slide dammed the North Fork of the Virgin River, destroyed a section of the road, and stranded some tourists, forcing them to extend their stay a few days. Mass wasting is the term used to describe the gravity-driven downhill movement of rock and soil, and landslides are one spectacular way in which this plays out. The landslide of 1995 occurred because the North Fork of the Virgin River had undercut an already unstable slope. To exacerbate the situation, 1995 was a year of unusually high precipitation. This combination of saturated soil and oversteepening allowed gravity to gain the upper hand, and a large section of the slope gave way. As you walk along the Watchman Trail, you can observe some other, less celebrated examples of mass wasting.

The trail begins along the North Fork of the Virgin River, but soon you climb up into the siltstone and sandstone of the Moenave formation. At this point, note the numerous large sandstone blocks scattered about. They are a result of rockfalls, one of the many forms mass wasting can take. The Moenave formation consists primarily of thin-bedded siltstone and sandstone capped by the more resistant Springdale sandstone

The Moenave formation is composed of alternating layers of sandstone and siltstone. The softer siltstone weathers into slopes, while the harder sandstone forms blocky ledges.

member. This formation is not the source of the majority of the larger blocks you see. They originated farther upsection, from the overlying Kayenta formation and Navajo sandstone.

The Kayenta formation is composed of alternating layers of siltstone and sandstone deposited primarily by rivers. The weaker siltstone layers in this formation erode much more easily than the sandstone, resulting in the characteristic pattern of small slopes and ledges this rock unit is known for. Over time, erosion of the siltstone layers begins to undermine and destabilize the overlying sandstone. Eventually, ever-patient gravity takes over and sections of rock weaken, break off, and tumble down. This process is ongoing, and small rockfalls are not unusual in Zion. As you wander, you may hear them or see the small dust clouds associated with them.

As you continue up the trail, you soon enter the slope and ledge terrain of the Kayenta formation. Here you can observe a slower form of mass wasting, referred to as creep, which occurs on almost all hillsides. Creep is the very slow, gravity-induced downhill motion of rock and soil. Though it may occur at an imperceptible speed, creep's effects are well known to quite a few homeowners. Many a concrete retaining wall has been cracked and slowly dismantled by this relentless process. Along this

As evidenced by this tilting tree, gravity is pulling this slope down, no matter how hard Dave Futey pushes.

section of the trail, you'll see another effect of creep: Trees naturally grow toward the sun, reaching skyward with tall, straight trunks. This is more difficult on an unstable slope, however. As gravity ever so slowly pulls the soil downslope, trees respond by slightly changing the angle at which they grow. This causes the lower trunk to curve upslope. Such trees with curved trunks are sometimes called drunken trees.

The end of the trail is a loop, and at the westernmost end of the loop is a nice view overlooking the town of Springdale. Look down to the north side of Utah 9, across from the cinema complex. The large hill with a few broken houses on it and a small café at its base along the road is an old landslide complex. It is composed of a jumbled mass of rock and soil deposited by numerous past landslides over thousands of years. This complex sits atop the Moenave formation, which is underlain by the Chinle formation. The latter contains a lot of claystone and mudstone and tends to be very unstable when it gets wet.

On September 2, 1992, an earthquake of magnitude 5.8 struck near the town of St. George, about 40 miles to the west (this earthquake is also discussed in vignette 7). Landslide deposits are inherently unstable,

The Springdale landslide block sits just above and to the right of the cinema. The scarp is the dark gray band above Utah 9.

Aerial photo of the Springdale landslide and scarp

and recent rains had saturated the soil, making a bad situation worse. The shaking associated with the earthquake caused a failure within the mudstone of the Chinle formation. This failure sent the upper part of the Chinle, the overlying Moenave formation, and the preexisting landslide deposits—46 million cubic feet of rock and soil in all—sliding over 30 feet downslope. Although this event is generally called the Springdale landslide, it was actually a specific type of mass wasting—a slump. A slump is a mass of rock and soil that moves downslope as a single unit, usually with some degree of backward rotation.

Toward the top of the slope, look for an irregular brown line (this is a good place to use binoculars if you have them). This is the main scarp along which the slump took place. It averages about 25 to 40 feet high. When the earth began to shake and the failure occurred, the whole mass of rock and soil below this line dropped down over 30 feet as a single unit. As the slump block slid down, it rotated backward slightly. A small

Deep fissures formed as the block broke apart during the landslide. This photograph was taken the day after the slide. Hiker for scale.

development had been built on the landslide complex, and the three buildings you see, along with the surrounding roads and infrastructure, were completely destroyed. Close inspection shows numerous fissures and minor scarps on the surface of the slump block. These indicate that the block did break apart slightly—at least at the surface—but it still traveled downslope as one coherent unit.

Note how the debris encroaches upon the highway. In the immediate aftermath of the 1992 event, this road was closed for a short time to allow for cleanup of rock and soil. Also notice that development in the area has proceeded unabated. Although, as of this writing, no more structures have been built directly on this unstable ground, construction has continued directly along its leading edge. The café on the north side of the highway, not yet built in 1992, seems particularly vulnerable to future downslope movement. As Springdale continues to grow, construction may continue in precarious locations simply because land suitable for building in this area is relatively scarce and there's money to be made in developments. This will put people and structures at risk from the processes described in this vignette: rockfalls, creep, slump, and landslides. This will put people at odds with something they can never quite control—gravity.

This home was destroyed as the slump block slid downhill.

11 A DESERT REBORN
Coral Pink Sand Dunes State Park

First-time visitors to deserts often expect to find a landscape covered with sand dunes. The reality is that sand covers only about 20 percent of the world's deserts and even less than that here on the Colorado Plateau. Some deserts in other parts of the world do feature exceptionally broad expanses of blown sand. The Sahara Desert, for example, is known as a sand sea, or erg, because of its huge accumulation of sand. Coral Pink Sand Dunes State Park, on the other hand, a much smaller region of blowing sand, is what geologists call a dune field. Some combination of wind, sediment, and topography has resulted in the sculpted forms that decorate this high valley.

In this vignette, we'll explore not only the types of dunes found here but also the underlying reasons for the existence of this colorful and unique dune field. From stop 1, you can see the southern portion of the dune field as well as the surrounding mountains. A few words of caution, however, should you decide to hike across the dunes: This park is a favorite of off-road vehicle (ORV) enthusiasts. Riders of these vehicles tend to fly over dune crests without any knowledge of what or who might be on the other side. Also, parts of the dune field are protected habitat for the Coral Pink Sand Dunes tiger beetle, a predatory insect that lives here and nowhere else in the world. If you see signs indicating protected habitat, please steer clear.

Although deserts experience little rain, moving water is still chief among the forces shaping their landforms. When precipitation comes, it often does so in torrents that race down hillsides, dive off cliffs, and fill previously dry channels, sweeping everything and everyone before

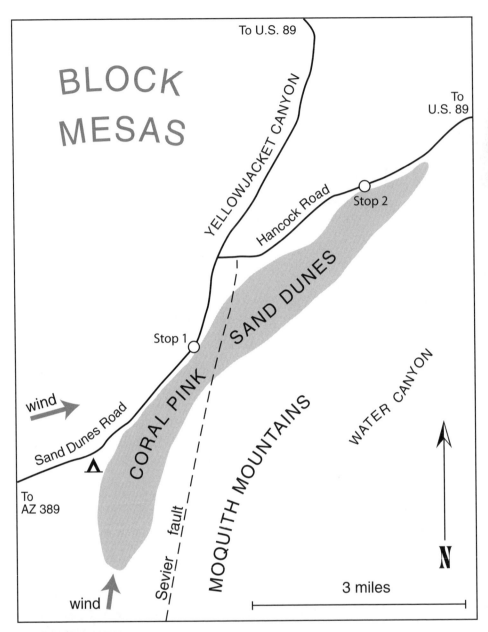

GETTING THERE

From Kanab, drive 14 miles north on U.S. 89, then south (left) on Sand Dunes Road. From Mt. Carmel Junction, drive 3 miles south on U.S. 89 and turn right on Sand Dunes Road. On Sand Dunes Road, it's 7 miles to Coral Pink Sand Dunes State Park. Stop 1 is the picnic area and viewpoint between the visitor center and the campground. Stop 2 is on Hancock Road just north of the park boundary. To get there, turn east off Sand Dunes Road onto Hancock Road and go about 2 miles for easiest access to the dunes. Park in any of the numerous pullouts. Note: This park is a haven for off-road vehicle enthusiasts, so the campground is prone to engine noise and fumes, especially on weekends.

The dunes at Coral Pink Sand Dunes State Park exist due to topography, wind, and the readily available sand grains from weathered sandstone.

them. But wind also plays an active, if secondary, role. The sparsity of water and vegetation, the latter of which normally holds soil in place and reduces wind stress, allows wind to be a more effective agent of erosion in deserts than in humid regions. And wind is most certainly the force behind the sweeping dunes here.

Coral Pink Sand Dunes State Park sits at an average elevation of 6,000 feet above sea level and covers 3,730 acres, about half the total area of the dune field. *Coral* refers to the deep pink color of the sand, a color derived from iron oxides cementing the sandstones that make up the local bedrock. The dune field is approximately 6 miles long and 0.5 mile wide, and it's elongated in a northeasterly (downwind) direction. Prevailing winds have been a major factor in creating this shape. The park lies at latitude 37° north, and winds between latitudes 30° and 60° north blow predominantly from the southwest to the northeast. The orientation of bordering mountain ranges has also played a role here. The Moccasin Mountains lie to the southwest, and the Moquith Mountains lie to the east. The latter range is really an uplifted plateau, and the prominent cliff that faces the park is part of the Sevier fault, discussed in vignette 16. Between the Moccasins and the Moquiths is an opening called a wind gap.

When wind is forced through a narrow opening, its velocity rises, and wind's ability to carry sand is a function of velocity. When wind velocity rises, more sand can be moved, and when velocity falls, some of the sand that's been carried along is deposited. Accelerating winds pick up sand grains weathered from sandstone bedrock near the wind gap and carry them into this valley. Beyond the gap, the valley broadens, which lowers wind velocity and results in deposition of sand and creation of the dune field.

The southern portion of the dune field is oriented more directly north-south than the northern end, which is oriented more to the northeast. Due west, just north of the Moccasin Mountains, is another wind gap. Winds forced through this gap from the west join winds blowing from the southern wind gap. The addition of these more westerly winds changes the orientation of the dune field. Not only that, the valley also begins to constrict in the region where the dune field's orientation changes, causing winds to accelerate and forcing sand up and over the Sevier fault escarpment in a northeasterly direction.

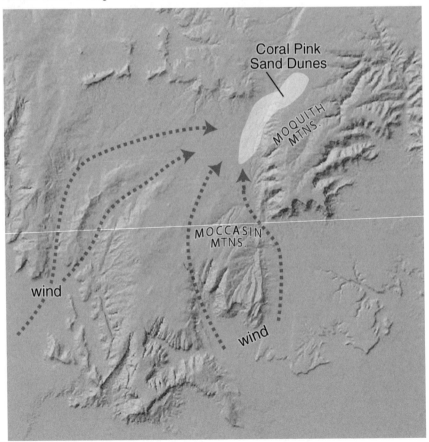

Topography influences wind flow at Coral Pink Sand Dunes.

These grasses act to stabilize small mounds of sand called coppice dunes.

Take a look at the dunes in front of you. Wind, not as powerful as water or ice due to its low density, can only move relatively small pieces of sediment. Silt and sand are both transported by wind, though the mode of transportation is different for the two. Silt particles are small enough to actually be swept into the atmosphere and suspended there by wind. Sand grains, on the other hand, are too large to be suspended in a column of air. They move instead by saltation, meaning that individual grains bounce across the surface in a downwind direction. Not all of the sand in this dune field is still moving. The southernmost tip of the dune field, as well as many of the margins, have been stabilized by vegetation. The interior, however, features four dominant dune types that are very recognizable.

If you look at a cross section of an isolated dune, you can see that it's asymmetrical. The windward face of the dune is less steep than the lee face. Saltating grains are driven up the windward face to the crest of the dune. There they cascade down the lee face, also called the slip face. The slope of the slip face is steeper near the crest than at the base. The slope at the base is called the angle of repose—it's the angle at which there's equilibrium between the downward pull of gravity and the frictional resistance between sand grains. Microavalanches are common on the unstable upper section of the slip face as the dune seeks equilibrium with gravity, and slip faces as a whole are only marginally stable; any

Microavalanches cascade down the slip face of a dune.

Dune types at Coral Pink Sand Dunes

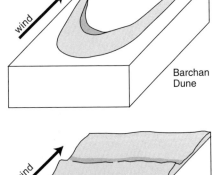

wind

Barchan
Dune

Transverse
Dunes

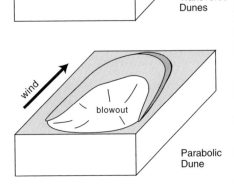

wind

blowout

Parabolic
Dune

disturbance will cause downward movement of sand. The slipping, sliding, and heavy breathing associated with hiking up the slip face of a dune certainly makes this clear. If you were to slice a dune in half vertically, you would discover that the interior of the dune is composed of fine layers called crossbeds, which represent the movement and deposition of sand on the slip face. In the long run, the net result of the movement of individual sand grains is the large-scale movement of the entire dune in the downwind direction.

The type of dune that develops from blowing sand is a function of both the sand supply and the prevailing winds. Four major dune types occur here at Coral Pink Sand Dunes State Park: barchan, barchanoid, transverse, and parabolic. Barchan dunes are crescent-shaped with tips that point downwind. They are solitary features that form when the supply of sand is limited and wind direction is constant. They migrate across flat stretches of land that have relatively hard surfaces and sparse vegetation. The orientation of barchan dunes here indicates that these dunes are being driven in a northeasterly direction. Where sand is more plentiful, barchan dunes coalesce to form barchanoid dunes, whose curved crests are roughly perpendicular to the prevailing winds. With an even greater supply of sand, transverse dunes form. Within a transverse dune field, linear crests and troughs are also oriented perpendicular to the prevailing winds. In the vicinity of stop 1, winds have constructed barchan, barchanoid, and transverse dunes. In order to see these three dune types, you will probably have to climb onto one of the higher crests. Do so carefully, ever mindful of the ORV menace.

Head to stop 2 to explore the northwesternmost part of the dune field and see the fourth type of dune. The broad parabolic dunes at stop 2 form where vegetation has stabilized part of the dune field. Wind is slowed by the patchy vegetation but still moves rapidly in open areas. The influence of vegetation produces high erosion zones called blowouts, large circular depressions from which wind has excavated sand. The end result is parabolic dunes with crescent arms that point upwind. Stands

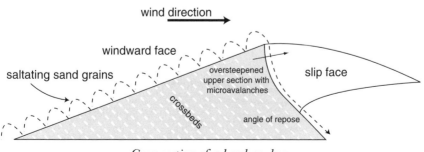

Cross section of a barchan dune

of tall ponderosa pines occupy the swales between dunes. If this area featured rapidly moving sand, saplings would be covered and killed before reaching the size seen here. Sand in this part of the dune field is not moving long distances; it's just being redistributed among stable vegetation communities.

There's an interesting interplay between bedrock and moving sand in this dune field. The lower portion of the Moquith Mountain cliff is Navajo sandstone from Jurassic time. This rock is what remains of an ancient erg that rivaled the modern Sahara Desert in scale, and every sand grain in the Navajo sandstone was once part of a dune (see vignette 8 for more on this topic). The Navajo sandstone is believed to be the source for much of the sand in the modern-day dune field here. Weathering of the ancient formation to the south releases individual grains, which the wind picks up and carries into this valley. So the dunes here are made of sand grains that are living a second life, exhumed by weathering and reanimated by wind. In areas around the western margin of the dune field, you can see living dunes adjacent to or resting upon outcrops of the lithified dunes that make up the Navajo sandstone.

The best time to visit Coral Pink Sand Dunes State Park, or any dune field for that matter, is in the early morning or in the evening when the sun is low in the sky. Painters and photographers have been drawn to the sensuous landscapes of desert dunes. Here in the park, the cool early morning air is redolent with the scent of juniper and sage, and strong shadows accentuate the sensuous forms of crests and arms. At such times, it's easy to understand the allure of wind and sand and also the desire to capture on canvas or film this constantly changing, yet always compelling, landscape.

12 CYCLES OF CHANGE
Castle Rock Campground

Castle Rock Campground, tucked away in the canyon of Joe Lott Creek, is more than just a pretty place to spend the night. The rocks here preserve a record of cycles of deposition, uplift, and erosion associated with the changing tectonics in this area. Joe Lott Creek is part of the Clear Creek drainage, which lies on the northern flank of the Tushar Mountains, which are in turn part of the extensive Marysvale volcanic field. Beginning about 30 million years ago, thick sheets of volcanic ash erupted out of five large calderas in the Marysvale field (see vignette 13). These were extremely explosive eruptions that sent high-speed clouds of glowing-hot ash, gas, and rock fragments out across the landscape. As the flow settled down, retained heat fused the ash and rock fragments into a type of rock known as a welded tuff. During the time the Marysvale field was active, southern Utah was relatively flat, and these flows covered hundreds of square miles, completely devastating everything in their path. One of these flows, the Joe Lott tuff, is exposed in this canyon; we'll discuss it a little later.

About 18 million years ago, this period of intense volcanic activity ended. Eruptions still occurred, but they were less frequent and on a smaller scale. Also at about this time, tectonic activity in the Great Basin began to stretch the crust, causing numerous faults to form. Large blocks of the earth's crust were uplifted and others were dropped down along these faults in a process that continues to this day. This faulting, combined with erosion, created a new drainage where Clear Creek now flows. Sediments began to erode off the higher areas and were deposited in the valleys, including the Clear Creek drainage, filling the canyons with layers of siltstone, sandstone, and conglomerate.

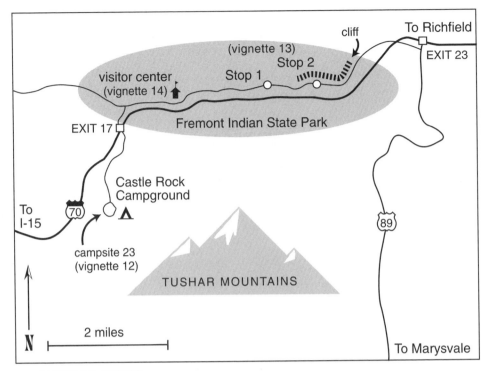

GETTING THERE

Castle Rock Campground, near Fremont Indian State Park, is on the south side of I-70, 18 miles east of the junction with I-15 and about 20 miles west of the town of Richfield. From I-70, take exit 17 and proceed 1.2 miles south down the campground road. Park in the day use area.

Here at Castle Rock Campground, remnants of these deposits, known today as the Sevier River formation, can be seen along the canyon walls. This formation consists primarily of tan to brown conglomerate and sandstone, and as you may have gathered from the hoodoos and spires everywhere around you, it weathers in a most extraordinary way. Walk around the campground loop and stop at campsite 23. Behind this campsite is a nice, easily accessible exposure of the Sevier River formation. Walk up to the cliff and examine the sediments. If you touch the cliff, you'll see that the cementation is poor; the rock tends to crumble quite easily. Note that different layers of sediment contain clasts—individual pieces of rock—of different sizes. These sediments were deposited by rivers, and during normal flow, a river carries only smaller, sand to pebble-size clasts like those that make up many layers of this formation. During storm events, however, flow increases and much larger clasts can be carried along. The size of some of the clasts poking out of the wall in front

Clasts in the Sevier River formation are almost exclusively volcanic.
The source of these clasts of ash flow tuff is the volcanic highlands to the south.

Intricate pattern of hoodoos in a drainage to the northwest

Castle Rock Campground got its name from the hoodoos weathering out of the Sevier River formation here. Note the two light-colored ash layers (higher one, top left). *The lower ash layer has been offset by small faults.*

of you makes it obvious that these streams sometimes moved with great speed and force. Layers of these larger clasts can be seen throughout the formation, and if you look at the ground around your feet, you'll see that many have weathered out of the cliff. They are almost exclusively of volcanic origin, indicating that they were carried down from the volcanic highlands to the south. After a storm, the stream returns to normal conditions and the deposits become finer grained once again.

The Sevier River formation is composed of basin-fill sediments, meaning that this formation didn't cover the whole region but instead was deposited in individual basins. These sediments are similar enough throughout the region to be included in the same formation, but each basin has its own history, and its sediments may differ slightly from other basins in the area. Note the two bright white layers, one in the lower third of the formation and the other closer to the top. These represent ashfall deposits that periodically blanketed the area—evidence that volcanic activity was still ongoing when the Sevier River formation was deposited. Note the difference in rock color above and below the lower ash layer. Beneath it, the color is a fairly consistent brown. Above the ash layer, however, the color is a much lighter shade of brown, gradually grading upward to the darker color. This tells us that ash was washed into the region's rivers and deposited downstream for quite some time after the actual eruption.

As with most rocks of volcanic origin, these ash deposits contain small quantities of radioactive elements and can be radiometrically dated (see vignette 16 for more on this method of dating). By obtaining dates for the two ash flows, we can infer the time frame during which the Sevier River formation was deposited. The lower ash bed is approximately 13.8 million years old, and the upper ash bed about 6.9 million years old. This tells us that sediments were accumulating in this basin from about 14 million years ago to just under 7 million years ago. Similar ash beds and basalt flows have been used to date the Sevier River formation in other basins, and the ages obtained have differed quite substantially.

The Great Basin is a restless place, and here on its eastern edge, faulting continued, uplifting the Clear Creek drainage. This uplift invigorated the stream, and it began to cut down into the Sevier River formation. Water, which had been an agent of deposition in this area, became an agent of erosion. The soft, poorly cemented rock of this formation offered little resistance to the erosive force of the water, so over the past 7 million years, much of it has been removed from this area. If you walk along the base of the cliff, you'll see numerous gullies that run vertically down the face, as well as some deeply incised channels. As water runs down from higher ground, it finds and exploits points of weakness in the rock and small channels begin to develop. Runoff from heavier rains is concentrated in these channels, in effect focusing the erosive power of the water not unlike a lens focusing light. Runoff from repeated downpours carves away at the soft rock, and the channels enlarge. Some of the deeper channels extend many feet into the cliff face. If you walk into one of them, you'll find that it ends at a vertical wall over which the runoff from the next downpour will come crashing down.

If you walk a short distance up the canyon, you can see evidence of the uplift of the Clear Creek drainage. The light-colored rock that underlies the Sevier River formation here is the Joe Lott tuff, one of the welded tuffs mentioned earlier. The Joe Lott tuff is about 19 million years old and originated from the Mount Belknap caldera, about 12 miles southeast of here. Its lighter, fine-grained matrix is composed primarily of ash, glass fragments, and small crystals of quartz and feldspar. The darker gray fragments are pieces of pumice.

Note that the contact between the Sevier River formation and the Joe Lott tuff is tilted quite substantially at this locality. In the area behind campsite 23, the Sevier River formation had a slight dip, but it was nothing like this (see vignette 6 for more about dip). The sediments could not have been originally deposited at an angle this steep, so this tilting is evidence of the faulting and folding that have occurred in this area. The whole block you're looking at has dropped down and tilted to the north along a normal fault located farther south along this drainage. A bit of diligent searching may turn up a few fist-size pieces of this unit that display smooth, polished surfaces. These polished surfaces, called slickensides, form as rocks on opposite sides of a fault grind past each other. Slickensides are considered definitive proof that faulting has occurred in an area.

Now climb a short distance up the outcrop for a view across the drainage to the east. Look at the relationship between the two rock units there: the Joe Lott tuff forms a nearly vertical cliff, with the beds of the Sevier River formation dipping away from it. The dip gets steeper and steeper as the beds approach the cliff, becoming nearly vertical at the cliff face. The contact between the Sevier River formation and the Joe Lott tuff is a fault scarp. As the block to the north dropped down along

Slickensides on a block of Joe Lott tuff from the fault zone

this scarp, friction along the fault surface bent the beds of the Sevier River formation upward. This bending of strata along a fault surface, called drag, is useful in determining the direction of motion along a fault. Along a normal fault like this one, the beds on the downthrown side (in this case, the Sevier River formation) bend upward and the beds on the upthrust side bend downward.

As you learn more about an area, you often begin to see things in the outcrops that you may not have noticed before. If you return to the outcrop of Sevier River formation behind campsite 23 armed with this new information about local faulting, you'll see that some of the beds in this formation are slightly offset due to faulting. This is easiest to see in the lower ash layer. The offsets aren't big, measuring only a few inches to a foot or so, but they are there nonetheless.

Sometime in the future, the rate of uplift may decrease. If that happens, downcutting will cease and the stream will start depositing sediment once again. At that point, the basin will start to fill again and a new cycle will have begun.

The Joe Lott tuff on the east side of Joe Lott Creek (left of photo) *has moved up relative to the Sevier River formation on the right. Drag along the fault has pulled Sevier River beds to a nearly vertical position.*

Middle cooling unit of the Joe Lott tuff. The poorly welded unit weathers to form rounded shapes.

GETTING THERE

Fremont Indian State Park is located on the north side of I-70, 18 miles east of the junction with I-15 and about 20 miles west of the town of Richfield (see map, p. 90). Take exit 17 off I-70 and follow signs to the park. The visitor center, 1.1 miles east of exit 17, is a good place to familiarize yourself with what the park has to offer. It has a fine collection of artifacts, including beautiful arrow and spear points. A short trail near the visitor center leads to petroglyphs. To get to stop 1, turn left out of the visitor center parking lot onto the frontage road and travel east 1.7 miles. Park in a large pullout on the left. To reach stop 2, continue east on the frontage road for 0.7 mile to a parking area on the right. Binoculars may be helpful, so bring them if you have them.

13 A RECORD OF EXPLOSIVE ERUPTIONS
Fremont Indian State Park

People have inhabited the canyon at Clear Creek in Fremont Indian State Park for thousands of years. The earliest inhabitants were hunters and gatherers who passed through and may have set up camp. Approximately 800 years ago, the Fremont Indians built a small village here where I-70 now passes; they also carved petroglyphs into the rock in this area. Though these petroglyphs have a tale to tell, what they say remains a mystery. The rocks into which these figures are carved also tell a story: They tell us that this part of Utah was not always as peaceful and serene as it appears today, and speak instead of violence and destruction. In this vignette, we'll help you understand their story.

Let's step back in time about 30 million years, when this area was part of a large belt of volcanic activity that stretched from the Great Basin of central Nevada to the western edge of the Colorado Plateau. Large volcanoes erupted thick sheets of ash and lava across what was for the most part a rather flat plain. The eastern end of this belt, where you now stand, is known as the Marysvale volcanic field. At least five large calderas—huge volcanic craters formed by immense, complex eruptions—were active in this area between about 30 million and 19 million years ago, and much of the rock exposed in Fremont Indian State Park and the surrounding area originated from these calderas. Lower levels of volcanic activity continued here until about 2 million years ago, but later eruptions were on a much smaller scale.

Go to stop 1, about 1.7 miles east of the visitor center, and look up at the canyon walls. The light-colored rock that comprises the walls of this small drainage is called the Joe Lott tuff, named after an early settler, Joe

Lott, who built a cabin on the Clear Creek floodplain in the 1870s. Tuff is a general term that refers to any rock formed from material ejected by an explosive volcano. But what you see here is a certain kind of tuff. The Joe Lott tuff is an ignimbrite (pronounced ig-NIM-brite), or ash-flow tuff, the product of an ash flow that contained a great deal of pumice. Ignimbrites are created during a particularly violent style of volcanic eruption similar to the 1980 Mount St. Helens eruption but on a more grandiose scale. We tend to think of the Mount St. Helens eruption in May 1980 as extreme, but at that time a mere 0.15 cubic mile of magma and debris was ejected from the crater. The eruption that deposited the Joe Lott tuff spewed out nearly 100 cubic miles of ash, almost seven hundred times the volume ejected from Mount St. Helens.

The Joe Lott tuff is one unit in a series of flows known as the Mount Belknap volcanics. These flows had two major source areas: one in the eastern part of the field in the southern Antelope Range, and the other to the west in the central Tushar Mountains. The Joe Lott tuff was deposited during an eruption of the Mount Belknap caldera in the western source area, about 12 miles southeast of the park.

There are many different types of volcanic activity, ranging from more or less passive emission of lava to explosive eruptions. Magma composition is one important factor in determining how explosive an eruption will be. Magmas rich in silica tend to erupt much more violently than basaltic magmas, especially during larger-scale events. The hands-down winner, however, in determining the explosiveness of a volcanic eruption is the amount of gases, also called volatiles, contained within the magma. When magma is deep beneath the earth's surface, the pressure of the overlying rock keeps gases in solution. As magma rises toward the surface, pressure decreases and the gases expand rapidly. The higher the percentage of gas in a magma, the bigger the bang.

Magmas that form ignimbrites are very rich in gases and hence very explosive, sending forth hot, glowing, churning clouds of ash and rock fragments that can completely devastate an entire landscape. These high-speed ash clouds spread out laterally across the terrain, so ignimbrites tend to be thickest in preexisting valleys and form only a thin veneer on the higher ridges. This contrasts with ashfall deposits, in which the ash falls like snow, covering the landscape to a similar thickness everywhere.

Ignimbrites are born of magmas high in silica, and one of their major components is pumice. If you're searching for evidence of gas in a magma source, you need to look no further than this bubble-filled rock. The small bubbles, called vesicles, form when dissolved gases come out of solution as magma rises to the surface. In some magmas, the gases can escape easily, and the resulting rock has few vesicles. In other magmas, such as

This cross section of ignimbrites is across the interstate from stop 1. The strongly welded lower unit forms the lower cliff. The top of this unit is poorly welded and forms the slope leading up to the middle unit. The upper unit caps the ridge. The pink unit can't be seen in this view.

the ones that form ignimbrites, the gas is still coming out of solution as the magma erupts. When this happens, the bubbles get trapped in the lava as it cools, forming highly vesicular pumice.

If you've ever seen a glassblower at work, you know that hot glass is soft and easily deformed. When an ignimbrite is erupted, the hot, foaming mass of silica-rich magma breaks apart into pumice fragments and small shards of glass. When the steaming flow finally comes to a stop, it settles and compacts. At this point, the pumice fragments, still hot and soft, flatten out under the weight of the overlying mass and fuse with the glass shards to form a much more coherent rock known as welded tuff. The most densely welded ignimbrites consist almost exclusively of glass and crystals. More commonly, though, ignimbrites weld only partially. Walk up to an outcrop here at stop 1 and examine the rock. The gray vesicular fragments are pieces of pumice. The light-colored, finer-grained matrix surrounding them is made up of glass shards and small crystals, mostly of quartz and feldspar. There may also be some darker black or brown bits. These are lithics, pieces of rock that were ripped from the wall of the vent during the eruption and went along for the ride. The rock is relatively light and chalky, and the pumice fragments

A close look at the middle cooling unit shows pumice fragments (gray) *and lithics* (black). *The pumice exhibits little flattening, which tells us the unit was only slightly welded. Hammer handle for scale*

don't appear to be flattened much. This part of the ignimbrite has been only slightly welded.

In the cliff face across the interstate here at stop 1, we can view a vertical section of the ignimbrite and divide it into what are called cooling units. Each cooling unit represents a single flow. (Sometimes flows erupt in such rapid succession that they cool as a single unit, called a compound cooling unit.) The Joe Lott tuff is composed of four cooling units, the lower, middle, pink, and upper units. The lower unit, at 200 feet thick, is the thickest of the four and is gray to light purple in color. The top and bottom of this unit are poorly welded, but the middle shows much stronger welding and in some areas displays beautiful columnar jointing. The lower unit makes up the sheer cliff face at the bottom of the section. Note that the upper, poorly welded section of this unit is lighter in color. We'll discuss this unit in greater detail at stop 2, where the columnar jointing is spectacular.

The middle cooling unit is about 140 feet thick and light gray to white in color. It consists predominantly of fine-grained material, with some light gray pumice fragments that are often concentrated in zones and darker lithics that tend to be pebble-size or smaller. This unit is poorly welded and thus erodes easily, sometimes forming interesting shapes.

*Dark lithics are concentrated in a layer that separates
the two ash flows of the pink cooling unit.*

There's some columnar jointing in the middle unit, but it's generally not as well developed as in the lower unit. In the view across the highway, this unit forms the slope and low cliff above the lower unit.

The pink cooling unit is about 85 feet thick and owes its name to its obvious pink color, derived from the oxidation of iron contained in the rock. This unit is actually composed of two ash-flow layers separated by a thin layer of ashfall. Light gray pumice fragments are found throughout the pink unit, but they're concentrated near the top. Dark lithics range in size from pebbles to small cobbles. This unit, like the middle unit, is poorly welded, and it weathers distinctively, with numerous cavities giving it a resemblance to Swiss cheese. The pink unit is difficult to see from here (binoculars help), but it is present atop the middle cooling unit.

The upper cooling unit, which caps the ridge across the highway, is about 100 feet thick. Recent exposures of fresh surfaces reveal that the unit is actually gray, but weathers to a tan or rusty color. Pumice fragments are numerous and found throughout the unit. Some of the pumice has been flattened, a clue that tells us this unit is moderately welded. This unit also contains more dark lithics than the underlying units.

Structurally, individual cooling units can be divided into distinct zones. The lowermost zone is a fine-grained, crystal-rich layer that is

The light-colored rock in the foreground at stop 1 is the middle cooling unit. The weathered rock in the center that resembles Swiss cheese is the pink cooling unit, and the upper cooling unit caps the cliff in the background.

at most only a few inches in thickness and may be completely absent. Next is a zone of fine-grained material that generally ranges in thickness from a few inches to a few feet. This fine-grained layer grades upward into a third zone, consisting of poorly sorted ash, crystals, lithics, and pumice fragments, the latter of which may be concentrated near the top. This zone makes up the bulk of the unit. Very fine-grained ash deposits sometimes cap the cooling unit, but these are rarely preserved.

Turn your attention back to the walls of the small drainage surrounding you for a closer view of a few of the cooling units we've been discussing. The light gray rock at ground level is the middle cooling unit. This unit forms some poorly developed columnar joints along this road, east of the parking area. Further up into the drainage lie the pink unit and the upper unit. Take a close look at an outcrop of the middle unit. The light-colored matrix is composed of ash, glass, and crystals—mostly feldspar and quartz. The darker gray fragments are pieces of pumice. Note the vesicular nature of these fragments. The small, dark pebbles are lithics. Near the top of this unit is a concentrated zone of pumice. Being very light, pumice tends to float to the top of a flow and can often be found concentrated there.

Weathering of ignimbrites is tied to degree of welding. The poorly welded pink unit (bottom) *weathers more rapidly than the upper unit* (top), *which is a moderately welded cliff former.*

Now look up at the pink unit. You can clearly see the thin ash layer separating the two flows that comprise this unit. The ash is gray and contains dark lithics that form distinct layers. Also note the numerous cavities eroded into the surface of the pink cooling unit, giving it the Swiss cheese appearance characteristic of this unit. The lithics tend to be slightly larger than those in the middle unit. If you look at a weathered block of the pink unit that has fallen or find a route up to the outcrop, you'll see that this unit is loaded with grayish pumice. The pumice fragments don't appear to be flattened to any degree, and this, along with the soft nature of the rock, tells us that this unit was only slightly welded.

The tan- to rust-colored rock that caps the walls of this drainage is the upper cooling unit. Since, as mentioned earlier, the upper unit is moderately welded, it's stronger and tends to form cliffs. If you find any blocks of rocky debris in the drainage in which the pumice fragments are flattened, they will have come from this unit. Both the pink and the upper units crop out at road level near the western entrance to the park.

After thorough exploration of the units exposed here at stop 1, proceed 0.7 mile east to stop 2, where the lower cooling unit displays very

The moderately welded lower cooling unit forms beautiful cliffs of columnar joints.

showy columnar joints. Step back and look at the cliff face looming above you. Although the individual columns are not quite as regular as those found at Devils Tower or Devils Postpile, this is still a fine example of columnar jointing. Two main conditions are required for formation of good columnar joints: the flow must be relatively thick, and it must cool relatively slowly. Other factors also play a role, including the homogeneity of the flow, the amount of air vesicles, and the presence or absence of internal layers. Columnar joints are most common in basalt flows, but as you can clearly see, they also form in silica-rich flows like the one before you.

The lower cooling unit is about 200 feet thick, so the first condition necessary for formation of columnar jointing—a relatively thick flow—has been met. As mentioned above, the upper and lower portions of this unit are only weakly welded. This is a consequence of relatively rapid cooling, wherein the flow cooled before it had a chance to weld effectively. Columnar jointing is not present in either of these parts of the flow. The middle portion, being more insulated, cooled much more slowly, allowing it to weld more strongly. This slower cooling also allowed the columnar joints to form in this section. If you walk across the frontage road and look at a block of talus near the base of the cliff, you can easily identify flattened pumice fragments, an indication that this part of the unit is moderately welded.

When rock cools—or any other substance, for that matter—its molecules become less active. As molecular activity decreases, the molecules themselves take up less space, so the rock contracts slightly. Columnar joints are cracks that form from stresses associated with this contraction. They form perpendicular to the cooling surface; in this case, the cooling surface was horizontal, leading to vertical fractures. This isn't always the case; for example, when a flow is steeply inclined, the joints can be almost horizontal. Columnar joints don't form all at once, but instead grow slowly and incrementally as the rock cools. As a fracture grows, it eventually comes in contact with other cracks. The long vertical fractures in this cliff formed in this manner. Over time, erosion exposed and widened these cracks, affording you this stunning view into the interior of the flow.

Not long after the eruption of the Mount Belknap caldera, the geologic regime in the Marysvale volcanic field began to change. The earth's crust in this region and to the west began to pull apart. This crustal extension resulted in formation of the numerous parallel block-faulted mountain ranges of the Basin and Range. The change in tectonics affected volcanic activity in the Marysvale field and brought an end to the large ignimbrite flows. Volcanic activity continued, but it was on a smaller scale and much less frequent. Some of these later eruptions took the form of localized basalt flows, while others sent clouds of ash into the air, which gently rained down on the landscape.

Later still, a drainage developed here and cottonwood trees grew. Cottonwoods mean water, a fact that the Fremont Indians understood well. In this land of abundant water, they occasionally found time to carve pictures into the rock. The petroglyphs remain secure in their ambiguity. The rock into which they were carved is less secretive. It clearly tells the story of cataclysmic events that occurred here so long ago.

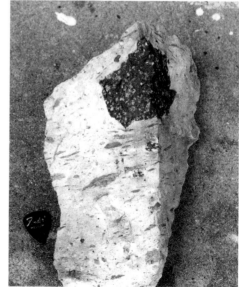

A dark lithic in a moderately welded block of the lower cooling unit. Note the flattened pumice fragments. Guitar pick for scale.

A nearly complete obsidian point and a broken one, both found in southern Utah

GETTING THERE

Fremont Indian State Park is located on the north side of I-70, 18 miles east of the junction with I-15 and about 20 miles west of the town of Richfield (see map, p. 90). Take exit 17 off I-70 and follow signs to the park. The visitor center, 1.1 miles east of exit 17, is a good place to familiarize yourself with what the park has to offer. It has a fine collection of artifacts, including beautiful arrow and spear points. A short trail near the visitor center leads to petroglyphs.

14 TOOLMAKER'S DELIGHT
Obsidian at Fremont Indian State Park

It can feel almost like magic when you're out walking and stumble upon a well-formed arrow point resting gently on the soil surface. Instantly your mind begins to wander; the object at your feet holds so many questions, so many mysteries. If you were to take that point and compare it to others found in the same area, you would find that they are, for the most part, composed of the same type of rock. Here at the museum in Fremont Indian State Park's visitor center, all of the points were found in this area, and all were made from a dark rock known as obsidian. Why was this the chosen raw material for all the ancient toolmakers in this area?

Each individual mineral has a definite chemical composition or a range of compositions, as well as distinctive properties that reflect its characteristic molecular structure. Physical properties such as color, hardness, crystal structure, density, and cleavage can be used to help identify minerals. For some minerals, color is a fundamental property that is constant; for others, color may vary. For example, malachite is always a distinctive green, whereas quartz, which is colorless when pure, takes on different colors when various impurities are present. Hardness, which basically means the ease or difficulty with which a mineral is scratched, is a physical property that's constant for any given mineral.

A mineral that breaks along certain planes is said to have cleavage. As with color and hardness, cleavage is a property that can be used to help identify a mineral. It's determined by the internal crystal structure of the mineral. Cleavage may be perfect, excellent, good, fair, or not present at all. Biotite is an example of a mineral with perfect cleavage in one

A chunk of obsidian displays a glassy surface and evidence of conchoidal fracturing.

direction, allowing it to be split into very thin sheets. The way a mineral breaks when it doesn't show cleavage is its fracture. Some minerals, such as quartz, display a special type of fracture referred to as conchoidal, in which a smooth, curved surface is formed. This type of fracture is the key to why native peoples often chose various forms of quartz, such as flint, chert, and chalcedony, to make tools and weapons.

Chemically, quartz is a very simple mineral consisting of silicon and oxygen atoms, which bond together to form tetrahedra, four-sided structures. A piece of quartz picked up from the desert floor is really a three-dimensional network of billions of such tetrahedra linked together. This sort of repeated linking of the same molecular group (in this case, the tetrahedron) is a process known as polymerization. Polymers formed by linking simple carbon-based molecules are the basis of all plastics and synthetic fibers.

Take a look at the large block of obsidian in the visitor center's museum. Note its glassy texture. This texture is the result of a combination of two factors. One of these factors is the rate of cooling. In general, the faster magma cools, the smaller the crystals that form. No matter how closely you look at this black chunk of rock, you won't be able to see any crystals in it. Obsidian forms from magma high in silica. When silica is heated above its melting point and then cools very rapidly, an amorphous solid results, commonly known as glass. Obsidian is a natural glass that occurs when a silica-rich magma cools so quickly that crystals don't have a chance to form. The other factor in obsidian's glassy texture is the degree of polymerization. Silica-rich magmas contain networks of linked silica tetrahedra, similar to those found in quartz. These magmas tend to be highly polymerized, and the higher the degree of

This obsidian flake displays the concentric pattern of a conchoidal fracture. Guitar pick for scale.

polymerization, the more easily glass forms. This polymerization is why thick, silica-rich flows can be glassy from top to bottom. These magmas are also very viscous and tend to ooze more than flow.

The high silica content and glassy nature of obsidian causes it to break in a conchoidal fracture, similar to flint or chert. The Fremont people who lived in this area 800 years ago may not have known about silica tetrahedra or polymerization, but they did understand that delivering a blow with the base of a deer antler would cause rock from certain source areas to fracture in this characteristic and highly useful way. Knowing and recognizing which types of rock to use and where to find them took an understanding of some basic geology, something the Fremont undoubtedly possessed.

Near the visitor center, you can see petroglyphs carved into a type of rock called ignimbrite, or welded tuff (see vignette 13 for more on ignimbrite). Like obsidian, ignimbrite is of volcanic origin, but instead of oozing slowly from a vent, it blasted violently out of a crater. Ignimbrites, composed of ash and pumice fragments, do not fracture to form usable stone tools. Thus, the obsidian used to make the arrow and spear points you see here in the visitor center, which were found in this area, had to have originated elsewhere.

The chemical characteristics of obsidian are unique for each individual source area. By analyzing known source areas and then comparing the characteristics of obsidian points to these sources, we can figure out where the obsidian used in the points originated. This information is also helpful in mapping ancient trade routes. When pieces of obsidian found here in the Five Finger Ridge area were analyzed, it turned out they came from two source areas. The majority came from the Mineral

Mountains, about 30 miles southwest of here. The rest came from the Black Rock source area, about 30 miles to the northwest. These source areas are close enough that the early residents of Clear Creek Canyon probably traveled there to obtain obsidian for points and other tools.

The process by which a chunk of obsidian from the Mineral Mountains was transformed into one of the delicate points seen here at the park museum was not easy. It took time, patience, skill, and probably a smashed finger or two. The process of making tools from stone, known as flint knapping, is an art common to ancient peoples the world over. This art is actually enjoying something of a revival among craftspeople today. The first step in the process is to choose a suitable piece of rock without flaws or fractures. Spending a few hours working a piece of stone only to have it split along a tiny fracture is quite frustrating, and after

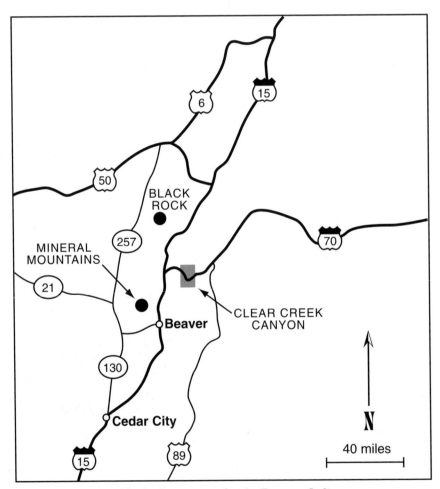

Possible obsidian sources for the Fremont Indians

it happens a few times, any serious flint knapper will become an expert at choosing good raw material.

Once a suitable specimen has been found, its initial shaping is done through a process known as direct percussion. In this process, the flint knapper directly strikes the stone with a tool to remove large flakes. Traditionally, the tool used for this purpose was either a waterworn cobble, called a hammerstone, that fit snugly in the hand, or a piece cut from the base of a deer or elk antler, known as an antler billet. By controlling the angle, placement, and weight of the blows, it's possible to fracture the stone in a predictable manner. Once the stone has acquired the general size and shape intended, the flint knapper refines it through another process—pressure flaking. In this process, a pointed tool such as an antler tine is placed on the edge of the stone, and pressure is applied to the tool. This pressure removes small, thin flakes from the edge, thus refining the point.

In many areas of the country, classes in flint knapping are available, or you can learn about it on the Internet. Give it a try sometime; you'll quickly develop a new appreciation for the skill required to form the exquisite points here at Fremont Indian State Park.

Outcrop of obsidian in the Mineral Mountains. Much of the obsidian used by the Fremont Indians in this area came from this source.

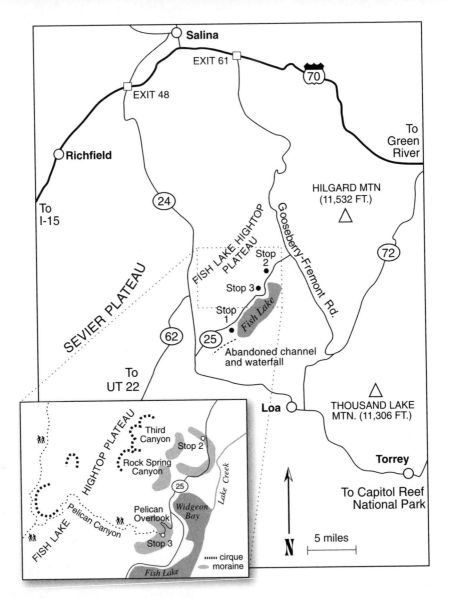

GETTING THERE

From the town of Richfield, go 8 miles east on I-70 to exit 48. Drive approximately 45 miles south on Utah 24 to Utah 25 and turn left (north). The southern end of Fish Lake is 6 miles farther on Utah 25. From the town of Green River, go 73 miles west on I-70 to exit 85, and take Utah 72 approximately 40 miles south to Utah 24. Turn right on Utah 24 for 12 miles to Utah 25, and turn right (north) to Fish Lake. Stop 1 is at the top of a steep rise in Utah 25 just before you reach the lake. Stops 2 and 3 are at the north end of Fish Lake. Stop 2 is a low ridge of glacial sediment on the left side of Utah 25 approximately 6 miles north of stop 1 and 2 miles north of the signed road to Pelican Overlook. From stop 2, turn back onto Utah 25 going south, then take the graded gravel road to Pelican Overlook. The turnoff to stop 3 is on the right about 0.5 mile up this road. The area surrounding Fish Lake is in Fish Lake National Forest.

15 ICE AGE GLACIERS
Fish Lake Valley

Fish Lake Valley, in south-central Utah, once hosted a number of alpine glaciers, flowing rivers of ice. During the colder climates of Pleistocene time, from 1.8 million to just 10,000 years ago, this moving ice shaped the landscape here and left behind clues to its existence. While looking at evidence of glaciation at Fish Lake, we'll discuss climate change and the dynamics of ice growth and movement. Fish Lake Hightop Plateau contains good examples of the types of glacial features scattered throughout

Fish Lake sits in a down-dropped block between two faults. Movement along these faults has altered the drainage direction from south to north.

southern Utah. Its glacial landforms are also some of the most accessible in the state. Here you can look at evidence of glaciation—including glacial erosion and deposition in close proximity—without climbing steep mountain trails.

Follow the directions in "Getting There" to stop 1 and safely pull over. A large streamcut valley lies to the southeast, yet no stream flows there today. An abandoned waterfall just south of the lake margin also indicates movement of a great deal of water sometime in the past. Today, water flows north out of Fish Lake into Lake Creek, continuing north into Johnson Reservoir and, in turn, the Fremont, Dirty Devil, and Colorado Rivers. Fish Lake Valley is a graben (pronounced GROBB-en), the German word for "grave." A graben is a down-dropped crustal block that forms a valley with faults bordering it on both sides. Geologists believe that movement along one or both of these faults about 10,000 years ago tipped the valley to the northeast, creating a sudden change in flow direction.

What evidence is there for a shift in the orientation of the valley floor? Fish Lake sits at an elevation of 8,405 feet above sea level, but the abandoned waterfall is about 8,445 feet above sea level—40 feet higher than the surface of the lake that formed it. Since water doesn't flow uphill, some change must have occurred to elevate the waterfall and channel in the southernmost valley. Although tectonic forces and faulting are not the focus of this vignette, they may have impacted some of the evidence for glaciation. We'll consider this further at stop 3.

About 1 mile north of Widgeon Bay (the northern extent of Fish Lake) and the road to Pelican Overlook, Utah 25 crosses a low, sinuous ridge on the valley floor. It crosses the ridge again 1 mile farther north at stop 2. Find a safe place to pull off the road near the northern ridge, and take a look at the roadcut. The ridge appears to be a jumble of boulders, cobbles, gravel, and dirt. Hike up onto the top of the ridge, which is about 8,500 feet above sea level. The surface is littered with boulders completely covered with lichen, a symbiotic community of algae and fungus that grows extremely slowly. In cold, dry climates, lichens may grow only 1 to 2 inches every thousand years. The fact that these rocks are so completely covered with lichen indicates that the ridgetop is an old surface, which is also confirmed by the low relief and smooth contours seen here.

So what are you standing on? This mile-wide, semicircular structure is a moraine, a pile of debris deposited by a glacier. If you look west at the flank of the Fish Lake Hightop Plateau, you'll see a deep valley with a curved head. Since this valley isn't named on topographic maps and it's the third canyon north of Widgeon Bay, we'll call it Third Canyon. The

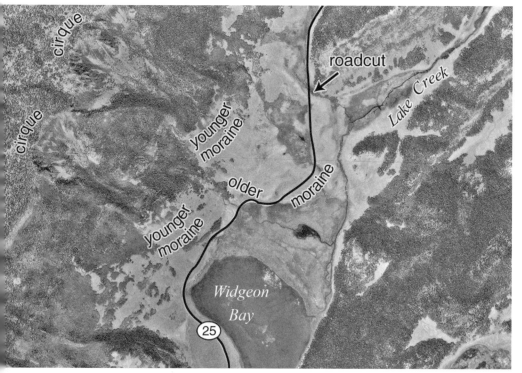

Aerial photo of Fish Lake moraines

base of Third Canyon is choked by a huge mound of debris—another moraine, but one much younger than the one you're standing on.

If you stroll west along the ridge crest toward Fish Lake Hightop Plateau, you'll encounter some trenches excavated by archaeologists looking for evidence of the Fremont culture, which occupied the shores of Fish Lake in the distant past. Without disturbing the trenches, peer in. You'll see the same jumble of rock and dirt you saw at the roadcut. This type of deposit, called till, is laid down directly by glacial ice.

Agents of erosion—water, wind, and ice—pick up, transport, and deposit sediment, in the process leaving their own fingerprint on the material. Streams, for example, create bedded deposits with uniform grain sizes, so sorting and stratification are the hallmarks of flowing water. Since glacial ice is solid, its dynamics of erosion, transportation, and deposition differ vastly from those of streams. Like a natural conveyor belt, a glacier carries an immense quantity of sediment out of highlands and dumps it all together in warmer lowlands, where the ice melts. These deposits lack sorting and stratification.

Drive back south to stop 3, Pelican Overlook. You can see Pelican Canyon on the way up, the southernmost of the two deeply incised

Exposures near Pelican Overlook exhibit the unsorted, unstratified structure characteristic of glacial till.

valleys south of Third Canyon on the plateau's eastern flank. The signed overlook turnoff leads to a graded dirt road that winds up onto the moraine at the base of Pelican Canyon. Examine rock exposures as you approach the crest. They should look familiar—the jumbled rock and dirt of glacial till. This is a moraine, but younger than the one at stop 2. The overlook, on top of the moraine at an elevation of 9,000 feet, offers a great view of Fish Lake and the down-dropped valley in which it sits. It also affords a good view west up Pelican Canyon, where the flowing ice of glaciers eroded huge quantities of bedrock then deposited it in this mound beneath your feet.

When did ice fill this valley? During Pleistocene time, commonly referred to as the ice ages, much of the earth experienced periodic changes from a cooler climate to a warmer climate then back again. In cooler, wetter periods, called glacial stages, continental ice sheets and alpine glaciers covered huge areas and lowered global sea level. More water in the form of ice on land meant less water in ocean basins. Warmer periods, or interglacial stages, saw the retreat of the ice and rising sea levels. Were today's ice sheets—in Greenland and Antarctica—to melt, sea level would rise well over 200 feet, inundating all of the world's port cities.

*The moraine at the base of Pelican Canyon has high relief
and is cut by narrow, steep-sided channels.*

The moraines you see here haven't been dated, but the prominent ones at the valley mouths have high topographic relief and are cut by narrow, steep-sided stream channels. They look very similar to other moraines in the Rocky Mountain and Great Basin ranges that are 20,000 years old. The older moraine at stop 2 has low relief and is dissected by broad channels—evidence of age. It is probably between 100,000 and 50,000 years old, based on its similarity to moraines of that age elsewhere in the Southwest.

Although there are no signs of older moraines associated with Pelican Canyon and its neighbor to the north, Rock Spring Canyon, the similarity of the recent moraines suggests that ice probably formed in all three canyons in earlier glacial stages as well. So older moraines probably existed in Pelican and Rock Spring Canyons. Perhaps the peninsula separating Widgeon Bay from the southern portion of Fish Lake is a remnant of a severely eroded moraine produced by an ancient glacier in Pelican Canyon. The tectonic shift in drainage may have caused surging northerly flows that eroded the southern two moraines but missed the northern moraine. On our illustration on the next page, we put big question marks where those moraines may have been.

Why did the glaciers melt? The geometry of the earth's orbit is always changing. The planet's orbital path can grow more or less elliptical, its axial tilt can become more or less acute, and the timing of solstices and equinoxes can vary. Any of these factors can alter the timing and quantity of solar radiation striking the earth's surface, and thus affect its climate. Earth has experienced earlier periods of intense glaciation—for example, in Precambrian time—but they were so long ago that little clear evidence remains, and we don't know much about them.

There are two broad categories of glaciers: continental ice sheets and alpine glaciers. Continental ice sheets, like those in Antarctica and Greenland, are huge masses of ice that accumulate in a central highland and flow outward to cover large areas. The Antarctic ice sheet, for example, has a maximum depth of 14,000 feet (well over 2 miles!) and covers 5 million square miles. Similar ice sheets covered much of North America and Europe during Pleistocene glacial stages.

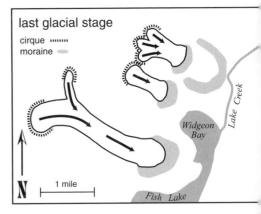

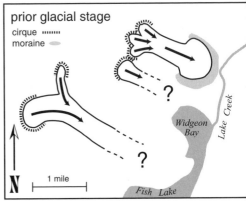

Reconstruction of past glacial stages

Alpine glaciers, on the other hand, flow in confined valleys. This type of glacier formed in the valleys of the Fish Lake Hightop Plateau and other mountains of the Southwest, while continental ice sheets developed to the north. Alpine glaciers form at high elevations, where more snow falls in a given year than melts. New snow layers compress each year's surplus, which eventually recrystallizes into a granular material called firn. Increasing pressure transforms firn into glacial ice, a pliable material capable of flowing and sliding downhill. Slap a ball of silly putty onto a sloping surface and watch it gradually flow to the bottom. Ice moves similarly, but over a much longer period of time.

Glaciers pick up sediment by a combination of plucking and abrasion. Meltwater often flows along the base of a glacier. This water frequently refreezes, essentially gluing rock on the valley floor to the base of the glacier. As the glacier moves, it tears up chunks of rock in a process called plucking. Incorporated into the base of the glacier, these pieces of rock behave like sandpaper grit, grinding underlying rock and removing additional fragments that the ice then picks up. Rock carried by glacial ice often bears striations—long scratches and grooves created by this intense abrasion.

As they erode bedrock, alpine glaciers produce distinct erosional landscape features. Whereas streams produce valleys that are V-shaped in cross section, glaciers carve broad U-shaped valleys. At the top of these valleys, amphitheater-like depressions called cirques mark where the head of the glacier lay. Two cirques are clearly visible at the head of Pelican Canyon, where ice formed in protected depressions to the north and south, then joined to form the Pelican Canyon glacier.

Eroded debris is carried to the lowest elevation at which glacial ice exists and is deposited in mounds as the ice melts. An end moraine forms at the toe, or terminus, of a glacier. The toe is rounded, resulting in a curved or arced mound. A lateral moraine, on the other hand, is material that was carried along the upper surface of the ice where the glacier met the valley walls. When the glacier melts, linear mounds remain along the valley sides. A third type of moraine, a ground moraine, is a horizontal layer of rubble deposited directly onto the valley floor during a glacier's retreat. Each period of relative climatic stability results in an end moraine. If the climate cools, a glacier advances to a new position of equilibrium and forms an end moraine at a lower elevation. If the climate warms, the glacier retreats and creates a new moraine at a higher elevation, or even melts entirely.

A trail up Pelican Canyon travels to the cirque floor (10,500 feet elevation), then up onto the plateau. It's about 3 miles through moderately steep terrain to reach the southern cirque. Hummocks along the trail represent minor periods of stability when the glacier sat at the same position long enough to build small debris mounds called recessional moraines. Between periods of stability, the glacier receded rapidly, depositing a ground moraine. If you choose to hike the trail, see if you can pick out these features. Also check for striations in the rocks that litter the trail.

When you look up from the trail, imagine a moving river of ice filling Pelican Canyon. Sniff the frigid air wafting down off the snow- and ice-covered plateau above you. Listen to sharp cracks as crevasses form in the ice and the muted rumbling of grinding rock from far below. Another, closer, rumbling reminds you that you are a Paleolithic nomad in search of dinner. You gaze hungrily at a woolly mammoth on the valley floor, then notice the massive dire wolf at the glacier's edge gazing hungrily at you. Welcome to Pleistocene southern Utah.

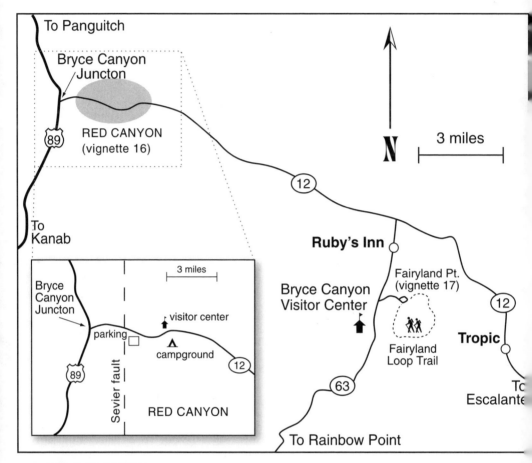

GETTING THERE

You can investigate the Sevier fault from a large paved parking area on the south side of Utah 12 at Red Canyon. Take U.S. 89 to its intersection with Utah 12, known as Bryce Canyon Junction. The junction is about 6 miles south of Panguitch and 59 miles north of Kanab. Go 2.3 miles east on Utah 12 to the parking area. From the east on Utah 12, the parking area is about 56 miles west of Escalante and 11.5 miles west of the intersection with Utah 63.

16 UNLIKELY NEIGHBORS AT RED CANYON
Sevier Fault

Here in southwestern Utah, in an area known as the high plateaus, the western edge of the Colorado Plateau is cut by three north-trending faults. They are, from east to west, the Paunsaugunt, Sevier (pronounced suh-VEER), and Hurricane faults. All three are normal faults, which means that the block above the fault, the hanging wall, has dropped down relative to the block below the fault, the footwall. In each of these faults, the western side has dropped down relative to the eastern side. Together these three faults form a transition zone between the Colorado Plateau and Basin and Range provinces. About 12 miles east of here, the Paunsaugunt fault forms the eastern edge of the Paunsaugunt Plateau and is responsible for much of the incredible scenery found at Bryce Canyon National Park (see vignette 17). To the west, the Hurricane fault (see vignette 7) forms the steep western boundary of the Markagunt Plateau, seen along I-15 on the drive up through Cedar City. The Hurricane fault is generally considered the western boundary of the Colorado Plateau. Here at Red Canyon, the Sevier fault forms the boundary between the Paunsaugunt Plateau, marked by the orange cliffs to the east, and the broad Sevier Valley, which opens to the west.

Though geologists can use numerous criteria to determine whether faulting has occurred in an area, generally only a few of these are present at any one locality. Some types of evidence are obvious, but other types take deeper insight and inference to discern. For your first clue that a fault is present here, look north across Utah 12 from the parking area. From right to left, you'll see an abrupt change, from the orange cliffs and spires of the Claron formation to black basalt.

View north from the parking lot. Quaternary basalt on the left (west) *side of the fault has dropped down about 200 feet and is now in contact with the older Tertiary Claron formation.*

The Claron formation dates from Paleocene to Eocene time and is generally divided into the Pink member below and the White member above. The deep reddish orange color of the Claron formation here tells us that the lower Pink member is exposed here. This rock unit consists predominantly of muddy limestone, calcareous sandstone, and siltstone, with lesser amounts of conglomerate. Take a short walk over to the large blocks of Claron at the east end of the parking area so you can get a closer look at this rock unit. Current theory holds that rivers slowly meandering across broad, flat plains deposited much of the Claron. Conglomeratic channel deposits, good indicators of deposition by rivers, are quite common here at Red Canyon. You can see large examples in U-shaped channels by hiking a short way into the ridges and cliffs north of the highway. Occasionally, momentous floods sent the ancient rivers over their banks, spreading mud and silt across the plains. Between floods, iron in the sediment was exposed to the atmosphere, and oxidation of this iron resulted in the deep reddish color of this rock.

If you look closely at these blocks, you may notice some vertical, V-shaped forms. These are root casts. Back in Eocene time, this sediment supported lush plant growth. When the plants died, secondary soil filled the space formerly occupied by their roots. The soil layers were then buried and subsequent lithification formed the rock you see here today, complete with the V-shaped root casts. These root casts are evidence that ancient soils, known as paleosols, were present here. Paleosols are

The Claron formation on the east side of the fault features brightly colored hoodoos like those found in Bryce Canyon National Park.

quite common in the lower Claron. Milky white crystals of the mineral calcite are visible in some of these root casts.

The black rock to the west of the pink cliffs is part of a basalt flow that erupted about 560,000 years ago. A large lobe of this flow is located to the west across Utah 12. When the strata of one formation, in this case the Claron, suddenly end next to another rock unit, such as the basalt here, it's possible that a fault may be present. This, however, is not definitive proof of faulting. The discontinuity may be the result of an unconformity—a gap in the rock record due to erosion. Or, more likely in this instance, it may result from an extrusive contact—a place where lava flows over preexisting rock. Mapping of this area has revealed that, to the east of the contact, this basalt flow actually sits on top of the Claron in some places. To the west, as you can see, the same flow lies at a much lower elevation. From this information, we can deduce that the contact seen here is actually a fault. The basalt has dropped down along the fault to become an unlikely neighbor of the Claron, and the northeast-trending line that separates the two is the fault line. This is the Sevier fault, and it has a total displacement of as much as 2,000 feet.

Shattered basalt at the contact

The change in rock type and color make the Sevier fault rather easy to recognize. In many instances, though, a fault doesn't have an obvious surface expression like this one, where the two rock units in contact are easily identified. Often the fault line is buried. In these cases, we use other, subtler clues associated with the change in rock type across a fault to infer faulting. Look to the north and follow the fault line. To the east of it stand the orange, jagged spires of Red Canyon. To the west lie the large, black mass of the basalt flow and the flat, open expanse of the Sevier Valley. As the fault block to the west dropped down, it left the higher block to the east exposed to erosion, which carved the cliff into the ridges and spires you see today. The western block, forming a topographic low, has not been subjected to these same erosive forces and remains relatively flat. Sometimes a sudden change in topography such as this indicates that a fault is present.

Now compare the vegetation on opposite sides of the fault line. The siltstone, sandstone, and muddy limestone of the Claron formation is somewhat crumbly and easily eroded and supports only a sparse piñon-

View south from north side of Utah 12. The fault line can be identified by the differing vegetation on the unstable Claron to the left versus the more stable basalt on the right.

juniper community. The basalt on the west side of the fault is much more resistant to erosion, and trees that take root here don't have the rock and soil constantly removed from beneath them. Consequently, tree growth is noticeably denser on the western side of the fault line in this area. This pattern is the same to the south.

Soils derived from different types of rock may vary in such factors as mineral content, pH, and soil stability. These variations affect what types of plants grow in a given area. A soil derived from limestone, for example, may have a higher pH and be better at holding water than a soil derived from sandstone. Therefore, plants that prefer high pH and need plenty of water will be able to thrive in soil derived from limestone but may not fare as well in soil derived from sandstone. Drought-tolerant plants, on the other hand, will find less competition in soil derived from sandstone and do better there. Sometimes such changes in vegetation can be used to map underlying contacts between rock units. This process is called biogeomorphology.

Here at the parking area, we can see an example of this relationship between rock and vegetation. It is relatively difficult to build a road across a basalt flow, so the engineers who built this section of Utah 12 smartly chose a route between two basalt flows. Follow the course of the road with

View west from the fault line. Utah 12 is at center left, a basalt flow lobe is at center right, and the Markagunt Plateau is on the horizon.

your eyes. Where the road crosses the area of the fault, the vegetation changes quite dramatically. To the east of the fault, the piñon-juniper forest grows on the Claron, but to the west, there's hardly a tree in sight. The open expanse of the Sevier Valley is covered almost exclusively in sage. Both piñon and juniper prefer the well-drained rocky or gravelly soil found on slopes. Sage, on the other hand, prefers wide, open spaces and can thrive in the more compacted soil of the Sevier Valley.

Often, upon discovering a well-exposed fault like this one, much time is initially spent observing and exploring the various relationships between the rock units. Before long, however, the question "Is it still active?" usually arises. In the Red Canyon area, the Sevier fault holds some information that may help determine the answer to this question. As mentioned earlier, the lava flow on the north side of Utah 12 occurs on both sides of the fault. If we can determine the age of the flow and then measure the amount by which it has been offset, we can figure out how much movement has occurred along the fault during that period of time.

We can assign absolute ages to some rocks using radiometric dating, a method that relies on the presence of naturally occurring radioactive elements. Volcanic rocks, such as the basalt at Red Canyon, form with small but detectable quantities of radioactive elements that decay naturally over time to become what are called stable daughter products. This decay occurs at a constant rate, denoted by an element's half-life, providing us with an atomic clock we can use to determine the age of a rock. (Half-life is the amount of time required for half of the radioactive parent isotope to decay to its daughter product.)

Initially the lava flow across the highway contained a certain amount of potassium-40, a radioactive element that decays to form the stable daughter product argon-40. The half-life for that particular process of decay is 1.3 billion years, and the atomic clock was set at the time the lava cooled and hardened to form basalt. The rock initially contained no argon-40, but over time, quantities of this isotope built up as potassium-40 decayed. By measuring the amounts of potassium-40 and argon-40 and then calculating the ratio of the two, we can calculate the age of this flow. If these quantities were the same, then exactly one half-life would have passed and the absolute age of the flow would be 1.3 billion

Investigating the contact between basalt and the Claron formation high above Utah 12

years. Any ratio between parent and daughter isotopes correlates to an exact amount of time between formation of the rock and the present. This technique has been used to estimate that the flow here at Red Canyon is about 560,000 years old. The offset of the flow from faulting has been determined to be about 200 feet, so in the last 560,000 years, about 200 feet of movement has occurred along this section of the Sevier fault. This substantial amount of offset of a relatively recent basalt flow suggests that the fault is probably still active.

At the beginning of this vignette, we mentioned that the Sevier fault is one of three large, north-trending faults that form a transition zone between the relatively stable Colorado Plateau and the heavily faulted and folded Basin and Range. The latter province, which encompasses western Utah, all of Nevada, and parts of five other western states and Mexico, has been subjected to tectonic extension for about 20 million years. This pulling apart of the crust has caused it to thin and fracture, and these three large faults in southwestern Utah are an eastern expression of this process. Given the substantial offset of relatively recent basalt flows crossing these faults and the fact that the tectonic extension that created the faults is still going on, it's likely that movement along these faults will shake up this region in the future.

17 A FAIRYLAND OF COLOR AND FORM
Bryce Canyon National Park

Geologists spend considerable time putting together sequences of events that occurred long ago. Because of this, people often get the idea that geologic processes occurred only in the past. There are places, though, that give a real sense of ongoing geologic processes—places where you can feel geology happening. Fairyland Point is one such place. Standing on the rim overlooking Fairyland Canyon, you can almost see the landscape melt away. The view to the east takes in a fantastic array of deep canyons, cliffs, fins, and spires, all in brilliant Technicolor. This landscape is constantly changing due to a combination of many geologic processes. Environmental factors, characteristics of the rocks themselves, and tectonic forces deep within the earth all continue to play a role in its development. Let's begin by taking a look at water and the key role it plays in shaping this ever-changing landscape.

The landforms here at Bryce Canyon owe their existence, in large part, to the erosive power of water. If you've traveled around the West, you may have noticed that rainfall, especially in the summer months, does not occur everywhere equally. You may be standing in one place in the sunshine watching a downpour happen not far away. In general, higher elevations receive more precipitation than lower elevations, and this holds true here on the Paunsaugunt Plateau, which averages about 6 inches more precipitation per year than the Tropic Valley, seen in the distance 2,000 feet below. The higher amount of rainfall at this elevation is important because the amphitheaters of Bryce aren't carved by runoff from the top of this plateau. The rim you're standing on is actually a drainage divide. All of the precipitation that falls behind you flows west into the Sevier River, which empties into the Great Basin. Only the water that falls directly on the slope in front of you plays a role in the creation of this fantastic scene. This water moves downslope through a system of intermittent streams that empty into the Paria River, which flows into the Colorado River, eventually ending up in the Gulf of California.

Erosion works better and faster when there is a greater change in elevation in a region. The vertical distance between the highest and lowest points in an area is referred to as relief. Here at Fairyland Point,

View east from Fairyland Point. The tall hoodoo with the dark top to the right of center is capped by channel conglomerate. The angled block in the background is the Sinking Ship.

GETTING THERE

To reach Bryce Canyon from the west, take U.S. 89 to its intersection with Utah 12, known as Bryce Canyon Junction (see map, p. 120). The junction is about 6 miles south of Panguitch and 59 miles north of Kanab. Proceed east on Utah 12 for 14 miles and turn right (south) on Utah 63. From the east, take Utah 12 out of Escalante, proceed 45 miles west, and turn left on Utah 63. Drive south for 3 miles and turn left on the Fairyland Loop Road, which is before the park entrance station. Fairyland Point and the Fairyland Loop trailhead are at the end of this short road. The Fairyland Loop Trail is 8 miles round-trip, but all the points discussed in this vignette are covered in the first 0.5 mile. If you do the whole trail, the elevation change is about 600 feet—somewhat strenuous but well worth the effort.

offset along the Paunsaugunt fault has created some serious relief. But the creation of Bryce Canyon is more complicated than a simple story of offset and ensuing stream erosion. In the view before you, just left of center, a block of tilted strata looks out of place in this landscape of relatively horizontal layers. This structure, known as the Sinking Ship, is thought to have resulted from drag along the Paunsaugunt fault, which lies just east of it. Fault drag occurs when layers of rock are bent up or down as a result of friction when movement occurs along a fault. The layers of rock in the Sinking Ship tilt down to the west. This tells us that the western block of the Paunsaugunt fault dropped down, dragging the strata of the Sinking Ship with it. So the Sinking Ship not only gives us an approximate location of the fault, it also tells us that you're standing on the down-dropped side of the fault and the Tropic Valley below you is on the uplifted side.

Here's how this apparent paradox came about: When movement along the Paunsaugunt fault began, the red rock unit you see here at Bryce, known as the Claron formation, covered much of this area, and there was little relief. Due to stresses originating far to the west, the earth's crust began to pull apart and a series of faults occurred in which the western side dropped down relative to the eastern side (see vignettes 7 and 16 for more on these faults). In the case of the Paunsaugunt fault, displacement has totaled about 2,000 feet. In the distance to the east, the shimmering, white Table Cliff Plateau, shows the elevation of the eastern fault block where erosion has not yet taken its toll. There the Claron caps the cliff at over 10,000 feet. Compare that to the 7,758-foot elevation of the Claron here at Fairyland Point.

The higher eastern block was then exposed to the ravages of erosion, which eventually peeled off the overlying Claron, exposing softer Cretaceous rocks underneath. Once those softer mudstone and sandstone units were exposed, downcutting began in earnest, wearing the eastern fault block down. If you walk far enough up any stream, you'll eventually reach a steep slope at the head of its valley known as a headwall. Rainwash, gullying, and slumping (the downward movement of saturated sediment) all work together to erode the headwall and lengthen the valley in the upstream direction. This process, called headward erosion, is particularly effective in young valleys. Headward erosion by the Paria River and its tributaries cut back into that young valley's headwall and formed the deep bowl of the Tropic Valley to the east. During this first episode of erosion, the lower block west of the fault remained unaffected. As time passed, though, the Tropic Valley widened, and east-flowing tributaries such as Fairyland Creek crossed the fault and began to cut back into the Claron formation on the fault's west side. This headward erosion into

Rapid erosion, typical of badlands, makes it difficult for this piñon to gain a lasting foothold.

the Claron formation west of the fault formed the colorful amphitheaters of Bryce Canyon National Park.

At Bryce, because the slopes are so steep and rainfall is relatively high, erosion occurs at a pace that plant life can't keep up with. As you walk along trails in the park, you'll see numerous trees desperately clinging to ground that's slipping away beneath them. Areas where erosion happens so fast that plants can't colonize the surface are called badlands. Many such areas take the form of a series of low, muddy, barren hills. This is obviously not the case at Bryce. Why?

The answer lies in the beautiful rock unit involved—the Claron formation. The Claron is composed of the lower Pink member (actually reddish orange), which you looked at earlier, and the upper White member, which is predominantly limestone. Erosion has removed the White member from this area, but it can be seen far off to the east at the top of the Table Cliff Plateau. It is the Pink member that has been carved into these fanciful hoodoos at Fairyland Point. If the Pink member consisted solely of layers of shale and siltstone, then the low, muddy hills typical of badlands would be the resulting landform. This rock unit, though, is much more diverse. While it does contain some easily eroded siltstone, it also has more resistant layers of limestone, sandstone, and conglomerate.

Conglomeratic sandstone caps this tall hoodoo along the Fairyland Loop Trail.

Water erodes these different rock types at different rates, a concept known as differential weathering. The shale and siltstone offer little resistance, and with each downpour the small intermittent streams fill with mud eroded from these beds. While the limestone in the Pink member also contains a lot of mud, it is more resistant than the siltstone. The limestone does eventually weaken, however, as rainwater, which is naturally slightly acidic, attacks carbonate in the rock. The sandstone and conglomerate are the most resistant to erosion, and they form the majority of the capstones on the hoodoos. Look out at the hoodoos in the amphitheater in front of you. Note that square blocks cap many of them. These capstones of resistant rock protect the column from erosion and hold it in place. Almost directly in the center of your view is a solitary hoodoo capped by gray conglomeratic sandstone. Now look at the cliff face to the left of it. A channel cut into the underlying siltstone is filled with the same gray sandstone that caps the hoodoo. As this cliff erodes back, more hoodoos with capstones of the same material will probably emerge.

From the overlook, you get a good view of this landscape and its strange sentinels, but to really feel the magic you must walk among them. A short hike down the Fairyland Loop Trail affords a closer look at examples of the principles we've been discussing. If you don't have

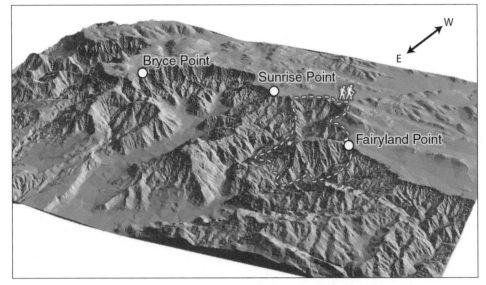

Three-dimensional image of the Fairyland Loop Trail

the time or inclination to hike the whole 8-mile trail, that's fine. All of our points will be discussed in the first 0.5 mile, after which there's a nice, flat, tree-covered area. Rest there and then come back up the way you went down.

The trail begins in thin-bedded siltstone. This rock erodes easily and becomes a supremely sticky muck when wet. As noted above, the Claron formation is composed of a variety of different rock types. This is a result of the environment in which it was deposited—large basins that contained meandering rivers and shallow lakes. Calcium carbonate accumulated in these shallow lakes, and also in soil in the form of caliche. Streams contributed sandstone and conglomerate. Occasionally these streams flooded, and as the floodwater receded, silt, the precursor of the siltstone seen here at the beginning of the trail, was deposited on the floodplain. The siltstone here erodes so quickly that plants can't gain a foothold. As you walk along this section of trail, look for small trees with exposed roots desperately trying to make a go of it.

As you continue down the trail, keep your eyes to the left, where you'll see a group of incipient hoodoos weathering out of a Claron fin. See if you can pick out the different rock types by their resistance to erosion. Each hoodoo has a blocky caprock, which in this case is muddy limestone. Directly under the caprock is a siltstone layer, which erodes more readily. Below that, in a fine example of differential weathering, a series of limestone, sandstone, and siltstone layers show varying degrees

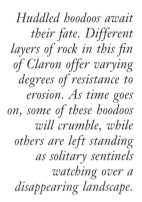

Huddled hoodoos await their fate. Different layers of rock in this fin of Claron offer varying degrees of resistance to erosion. As time goes on, some of these hoodoos will crumble, while others are left standing as solitary sentinels watching over a disappearing landscape.

of weathering. The resistant limestone cap holds the whole unit together. Continued erosion will cause some of these incipient hoodoos to crumble; others will remain standing, solitary sentinels guarding a disappearing landscape.

Another characteristic of the Claron formation that enhances hoodoo development is the presence of smooth fractures known as joints. Stresses due to tectonic activity or deep burial under other rock can build within sedimentary rocks. Folding and faulting are common mechanisms for releasing stress, but often some stress remains in the rock. Joints may form in response to this residual stress. Joints differ from faults in that there's no movement along the surface of joints. Joints aren't distributed randomly in rock, but rather are grouped together in more or less parallel sets. The Claron contains two sets of nearly vertical joints. One is oriented toward the northeast and the other toward the northwest.

Initially there's little space between joint surfaces, but over time, weathering may widen this space. We mentioned earlier that slightly acidic rainwater can dissolve limestone; this mode of chemical weathering can increase the space between joint surfaces in that type of rock. Another type of weathering, frost wedging, also plays a significant role. If you've spent much time on Utah's high plateaus, you know that no matter how warm it gets during the day, it still gets quite chilly at night. Temperatures can vary widely within a twenty-four-hour period in this region. Often it gets cold enough at night to freeze the water that seeps into joints and other small fractures. Unlike most fluids, water expands slightly when it freezes, and this expansion wedges open the fractures a little wider and a little deeper. The following day may warm above freezing, melting the ice. The water is then free to seep into fractures once again, penetrating slightly deeper. This cycle, repeated over and over, eventually breaks the rock apart.

Well-weathered vertical joints are prominently displayed in this outcrop. The face of the fin is oriented along one set of joints, and the hoodoos are separating from each other along another set. From certain angles you can see spaces beginning to form between some of the incipient hoodoos. Look out across the landscape and observe the orientation of other groups of hoodoos. They too will be oriented along these same two planes.

As you continue your walk, note the shape and composition of the various hoodoos you pass. Try to figure out what type of rock forms their capstones, when capstones are present. After about 0.5 mile, the trail levels out onto the tree-covered plateau we mentioned earlier. Find a comfortable spot, take a break, and look across the drainage at the hoodoos weathering out of the opposite side. Note their various shapes. Many have the blocky caps discussed above, but others are pointed at the top. The pointed hoodoos have lost their caprocks. Once its caprock is dislodged, a hoodoo erodes much more rapidly. Some of the hoodoos have resistant layers farther down the column. When such a layer is exposed, it can act as a caprock on the now more diminutive hoodoo. Other, less fortunate, hoodoos will be reduced to nubs relatively quickly. Apply your knowledge of the rock types in the Claron formation and try to pick out which hoodoos may have a second chance.

If you're feeling spry, you can continue around the loop and visit Tower Bridge, a natural bridge 3.5 miles ahead. The lighting changes throughout the day, making this a truly enchanting walk. Or you may prefer to remain here relaxing for a while and listening for falling pebbles or small avalanches—the sounds of geology in action.

Tower Bridge stands at the midpoint of Fairyland Loop Trail.

*The Fairyland Loop Trail introduces visitors to
a variety of forms and structures.*

*The process of erosion
continues whether we
build trails and
observation points
or not. This chunk of
the Claron formation
near Fairyland Point
will someday be
just a memory.*

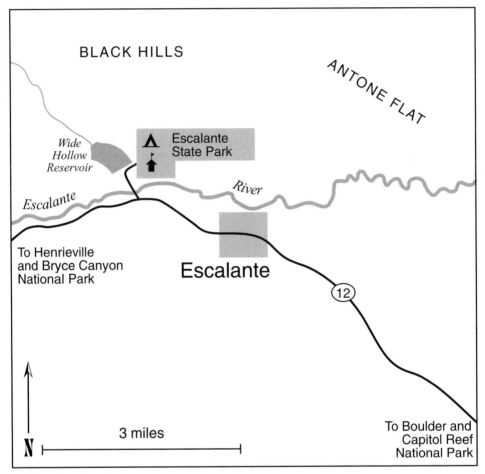

GETTING THERE

To reach Escalante State Park, take Utah 12 either 1.5 miles west of the town of Escalante or about 44 miles east of Bryce Canyon National Park. Turn north on North Reservoir Road and drive 0.5 mile, past Wide Hollow Reservoir. The park offers camping and a small but informative visitor center. There are two main trails within the park. The 1-mile Petrified Forest Trail ascends 250 feet along a ridge to a plateau. The 0.75-mile Trail of Sleeping Rainbows, a branch of this trail, is steeper but well worth the effort. If you'd rather not hike, the Petrified Wood Cove adjacent to the campground displays many large pieces of petrified wood.

18 NATURE'S PALETTE
Escalante State Park and the Petrified Forest

Nature's extravagant use of color is demonstrated time and again across southern Utah. It can be seen on a crisp March evening as the setting sun illuminates the salmon-colored Entrada sandstone of Delicate Arch, with the snowcapped La Sal Mountains in the background. You see it in the pink ripples of sand emerging from shadows at Coral Pink Sand Dunes State Park. In narrow reaches of Zion Canyon, tall cliffs of Navajo sandstone carved by the North Fork of the Virgin River are stained a

Colorful blocks of petrified wood decorate the trail.

139

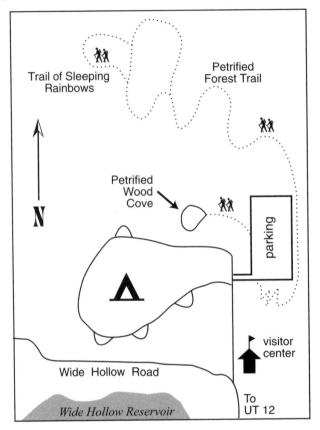

Trails in Escalante State Park

deep rust color and grudgingly give way to streams of light. Such colors are the result of millions of years of geologic processes. However, perhaps nowhere else in Utah are the colors of rock—or, more precisely in this instance, of fossilized deposits—more varied than in the petrified wood in Escalante State Park.

Petrified wood displays nature's artistry with a mineralized palette of colors. Though petrified wood has been found in a number of locations throughout the country and is perhaps best showcased at Petrified Forest National Park in Arizona, Utah's Escalante State Park displays a bountiful assemblage. The park, established in 1963, is estimated to contain nearly 5.5 million tons of petrified wood within its 1,400 acres, from tiny, fragmented chips to logs with cross sections 3 to 4 feet in diameter. There are several places in the park to see the wood, but hiking the park's trails allows you to see it in its correct geologic setting. The Petrified Forest Trail shows off many pieces of colorful wood, revealed as the substrate around them weathers and erodes away. The first section of this trail ascends 250 feet, but the majority of it is fairly flat, looping through the piñon-juniper woodlands atop the plateau. The real treasures can be found along the Trail of Sleeping Rainbows, a steep walk with many large logs. Whichever trail you choose, you'll see a variety of colors and forms.

Petrification is a process in which minerals replace the cellular structure of wood or other biological materials. Dinosaur bones, plants, and even ancient waterfowl eggs have been discovered in petrified form. Petrification can occur through three different processes: replacement, with minerals such as calcite and silicates taking the place of an original material such as wood; permineralization, in which minerals fill original pore spaces in an organic structure; and recrystallization, in which the original structure goes through a crystalline change and forms a stronger structure. Replacement with silicate minerals—in this case, agate—is the type of petrification that occurred here at Escalante State Park. Silicates, including opal, quartz, and feldspar, form the most common class of minerals and as such are widely found. The replacement process occurs at a chemical and microscopic level. It's influenced by several variables, including how quickly wood is buried (faster burial limits exposure to oxygen, which speeds decomposition); the types of cell structures and minerals involved; the duration of the process (a slow replacement may more clearly duplicate the wood's cell structure); the composition of the sediment the wood is embedded in; and influences related to temperature and moisture.

When silica replaces organic matter, it can preserve growth rings such as these, and even cell structure.

The really big logs, and lots of them, are found on the steep Trail of Sleeping Rainbows.

Channel deposits, fossils, and petrified wood characterize the Morrison formation, of latest Jurassic to earliest Cretaceous age. This formation, deposited 146 to 138 million years ago, dominates the park's exposed stratigraphy. Deposition of the Morrison formation was not limited to the Escalante area or Utah; it covered nearly 750,000 square miles of what is now the western United States. Here in the park, three members of the Morrison formation can be seen above the Escalante sandstone. From oldest to youngest, they are the Tidwell member, which consists of beds of red mudstone and white sandstone; the Salt Wash member, well exposed here, which forms cliffs consisting of white sandstone beds, conglomerate sandstone beds, and interbedded reddish mudstone; and the Brushy Basin member, composed of dark brown conglomerate, mudstone, and cross-stratified white sandstone. The white to light gray Escalante sandstone is exposed in the campground area and forms the base of the cliffs that make up the southern boundary of the park.

The Escalante sandstone is the modern-day descendant of windblown dunes created during the retreat of the Sundance Sea during middle Jurassic time. Influenced by the breakup of the supercontinent Pangaea, of which the continental United States was the western portion, the shallow Sundance Sea crept down from the north over much of what is now Utah, Wyoming, and Montana. The area around you was the sea's southern boundary, so you're standing on a 150-million-year-old beach.

Care for a swim? Later, as the sea retreated, southern winds spread the beach dunes over much of southeastern Utah. Had you been able to hike here at that time, you would have experienced a semiarid to arid climate on a landmass that was farther south, much closer to the equator. There were, however, periods of intense seasonal precipitation. That precipitation, along with streams from western highlands, sustained dinosaurs and other animals, aquatic life, and a variety of plants.

Over time, precipitation increased, creating extensive river systems that fanned across the basin. Periodic flooding and erosion uprooted trees, most likely conifers, though we can't be sure since their cell structure wasn't preserved. The uprooted trees were deposited and buried in sandbars and on floodplains between river channels. Toward the end of the episode recorded in the Morrison formation, volcanic activity deposited ash and tephra (pieces of volcanic rock ejected into the atmosphere) in the southern and western highlands, headwaters of the region's rivers. Within the river systems, these materials were deposited in floodplains and over riverbanks, covering the fallen trees with mud and silt, which became mudstone.

As the volcanic ash in the mudstone weathered, it released silica and other minerals into the water and muddy ground. With the trees firmly encased below, the mineral-rich silica solution slowly seeped downward, replacing each decaying wood cell and cavity. The other minerals in the silica solution account for the astounding colors you see so readily on the park trails. At sign 10 on the Petrified Forest Trail, the large recumbent log is a marvelous example of the spectrum of minerals and colors found in this park. Here are some of the relationships between minerals and the colors seen here: copper, cobalt, and chromium produce shades of

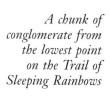

A chunk of conglomerate from the lowest point on the Trail of Sleeping Rainbows

blue and green; manganese produces pink; carbon and manganese oxide produce black; and iron produces a rainbow of colors on its own—red, brown, yellow, and green.

If you choose to hike the Trail of Sleeping Rainbows, you'll descend into a steep ravine, cross a wash, and walk up the other side back to the Petrified Forest Trail. On the Trail of Sleeping Rainbows, you'll see magnificent petrified logs with cross sections 3 to 4 feet in diameter. At the halfway point, the Bailey Wash overlook stands over a cliff. If you look beneath your feet and back at the hillside you just walked down, you can see exposures of conglomerate, a rock made of many rounded stream gravels deposited in channels and bars and now cemented together by silica, calcite, and iron oxide. This type of sedimentary rock reflects river action.

As you return to the trailhead, think about the many millions of years required to create a single piece of petrified wood, each one a unique work of art. Collecting petrified wood is prohibited in the park, so enjoy your visit, but leave the wood here for others to appreciate. If you don't, you may be tempting fate: the information display case at the trailhead includes letters from visitors who suffered mysterious maladies after taking petrified wood as souvenirs.

19 TRAVELING THROUGH TIME
The Waterpocket Fold,
Capitol Reef National Park

Waterpocket Fold. It just rolls off your tongue. Waterpocket Fold. Deep mystery lurks within these words. Say them in the evening when the light is just right and it will send chills down your spine. It may be the great white domes of the Navajo sandstone that give this park its name, but its heart and soul lie deep within the Waterpocket Fold.

Early explorers and mapmakers traveling the American West referred to impassable areas of rock as reefs. The Waterpocket Fold is one such barrier, exposing a nearly unbroken ridge of tilted rock almost 100 miles long. Massive sandstone units make up the backbone of this structure. Over time, erosion has sculpted holes and depressions in the sandstone. These depressions, or pockets, fill with water after rains or snowmelt, giving the feature its name.

The Waterpocket Fold is but one of the numerous, lengthy kinks in the bedrock of the Colorado Plateau. Known as monoclines, these areas of steeply tilted rock with relatively horizontal beds on either side are formed by compression; in the case of the Waterpocket Fold, the compression was caused by the collision of the North American Plate with the Pacific Plate in late Cretaceous and early Tertiary time. This compression caused a fracture, or fault, to form in the crystalline basement rocks deep below the earth's surface. Rocks on the west side of this fracture were uplifted as much as 7,000 feet relative to those on the east side. The thick layers of sedimentary rock overlying the crystalline basement reacted differently to this stress. Instead of fracturing, they bent to form a ramplike structure connecting the lower horizontal strata to the east to the higher strata to the west. Later, general uplift exposed

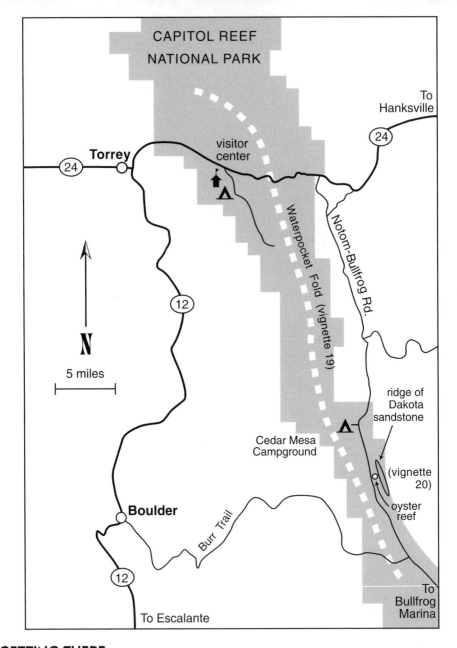

GETTING THERE

Capitol Reef National Park is located along Utah 24, 11 miles east of Torrey and 40 miles west of Hanksville. This vignette takes us from the visitor center at the western entrance to the east entrance, with several stops in between. Follow the signs to stop 1, the visitor center, about 7 miles from the park's west entrance and about 9 miles from the east entrance. Stop 2 is a parking area on the north side of the road 0.4 mile east of the visitor center on Utah 24. Stop 3 is the trailhead to Hickman Bridge, 1.5 miles farther east on Utah 24. The trail is less than 2 miles round-trip. From stop 3, drive 4.7 miles east on Utah 24 to stop 4, Fremont Falls. Stop 5 is a parking area 1.9 miles east on Utah 24, on the southern side of the road. Stop 6 is the parking area at the east entrance to the park.

these strata to intensified erosion and thousands of feet of rock were removed, bringing the internal structure of this magnificent monocline into full view.

Structure, however, is only one part of the story told here. A monocline like the Waterpocket Fold also gives us a unique opportunity to almost effortlessly examine thousands of feet of strata without having to scale cliffs or drive endless miles to various outcrops. Here at Capitol Reef, we can drive a mere 9 miles and see thousands of feet of rock belonging to numerous formations that were deposited over nearly 100 million years. Each one of these formations gives us a snapshot of what the environment of this area was like at the time of its deposition. By

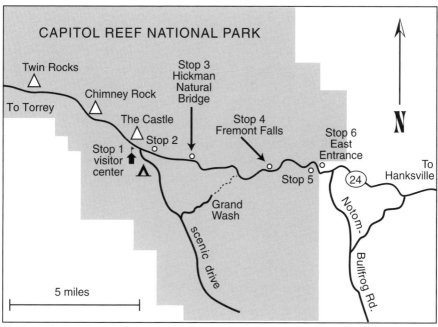

Stops along Utah 24 as it traverses the Waterpocket Fold

Monocline

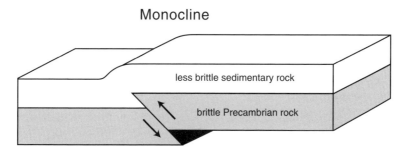

Basement rock fractures, while surface rock warps to create a monocline.

*View north from the visitor center. The dark rock at the base of the cliff
is the upper Moenkopi formation. The slope former in the middle is
the Chinle formation, overlain by Wingate sandstone.*

piecing these snapshots together, we can form a mosaic that represents
a substantial piece of the geologic history of this region.

Follow the directions in "Getting There" to stop 1, the visitor center,
where you'll begin your journey through time about 240 million years
ago, in early Triassic time. Older rocks are present to the west in the
deeper canyons, but they're difficult to get to, so we'll begin with the
first rock unit you can easily put your hands on.

From the parking area, look at the cliff across Utah 24 to the
north. Thin layers of siltstone and sandstone of various colors make up
much of this cliff, which is capped by The Castle, a feature carved out
of the massive Wingate sandstone. The thin-bedded, chocolate brown
rock unit at the base of the cliff is the Moody Canyon member of the
Moenkopi formation. This uppermost member of the Moenkopi formation is composed primarily of alternating layers of shale, siltstone, and
fine-grained sandstone. On the south side of the visitor center, a trail to
the campground parallels the scenic drive. Take a short walk along this
trail and immerse yourself in this part of the Moenkopi.

When the Moenkopi was being deposited during early Triassic time,
south-central Utah was part of a broad, flat plain with a shallow sea to

*Alternating shale and siltstone beds form the Moody Canyon
member of the Moenkopi formation. Hammer for scale.*

the west and the highlands of the Ancestral Rocky Mountains far to the
east. Slight fluctuations in sea level caused the shoreline to migrate back
and forth across this plain throughout the time this formation was being
deposited. During deposition of the Moody Canyon member, sea level
was relatively high and shallow marine to tidal flat conditions prevailed.
Note the layers of shale and siltstone that make up most of this unit.
These represent changes from quiet, shallow-water deposition (producing the shale) to an environment with slightly higher energy that could
transport larger particles (producing the siltstone). Some of the siltstone
beds display nice ripple marks, evidence of this higher energy environment. At the visitor center, many of the ripple-marked siltstone blocks
in the low walls bordering the sidewalks came from the Moenkopi.

Small pieces of gypsum are also abundant in this rock unit. They
are especially easy to find on clear days, when they glisten in the sunlight. Gypsum belongs to a group of minerals known as evaporites. In
arid regions, the rate of evaporation is very high. As a body of water
evaporates, minerals like gypsum precipitate out, sometimes forming
thick deposits. This gypsum, then, adds another piece to our picture of
south-central Utah at this time. It tells us that our early Triassic tidal
flat existed in an arid climate.

After thorough exploration of the upper Moenkopi, proceed to stop 2. At the base of the outcrop here, you'll see the now familiar brown siltstone of the Moody Canyon member of the Moenkopi. The point at which the color of this slope changes to light purple and gray marks this unit's contact with the overlying Chinle formation, of late Triassic age. This contact is an unconformity, meaning there's a gap in the rock record here, a gap that swallowed all evidence of middle Triassic time. Unconformities can simply represent periods of time when no deposition was occurring, or they can reflect times when erosion removed previously deposited sediments and rock units.

The Chinle formation in this area is composed of four members, although only three of them are present at stop 2. The lowest, the Shinarump conglomerate, was deposited in valleys cut into the underlying Moenkopi formation. Although it's missing in this outcrop, exposures do occur just to the west, forming the caprocks of such features as Chimney Rock and Twin Rocks. Here at stop 2, it is the next layer up, the colorful Monitor Butte member, that's in contact with the Moenkopi.

By late Triassic time, sea level had dropped and the shoreline was far to the west, in central Nevada. An intricate system of rivers carried sediments from the east and southeast across central Utah to empty into the sea. Rainfall increased, creating tropical conditions and promoting the growth of lush vegetation in marshes and around numerous lakes. The rocks of the Monitor Butte member reflect these changes. Walk a short distance along the base of the outcrop and examine the claystone that makes up much of this rock unit. Note the crumbly texture of the surface of the slope. At the time these sediments were accumulating, volcanic activity to the west periodically covered the landscape with ash. Subsequent weathering altered the ash to the clay mineral bentonite, which tends to absorb water and expand when it gets wet. When bentonite dries, it remains in this expanded form, so a crumbly surface texture such as you see here is a good indication that volcanic ash is present in a rock unit.

If you look around on the ground, you'll see some yellowish to cream-colored nodules scattered about. These carbonate nodules are thought to have formed in ancient soils, referred to as paleosols. At times during the formation of the Monitor Butte member, neither deposition nor erosion was occurring. Extensive soils developed at these times, supporting lush plant growth. Some of the Monitor Butte paleosols are riddled with trace fossils.

The Petrified Forest member overlies the Monitor Butte member in Capitol Reef. Meandering rivers that occasionally flooded deposited the siltstone and sandstone that make up this unit. The crumbly surface texture can be found here too, evidence that nearby volcanoes continued to eject ash at this time. Near the top of this unit, you'll see a resistant ledge of sandstone known as the Capitol Reef bed. This crossbedded sandstone was deposited in channels by fast-flowing rivers and represents a higher-energy environment than the siltstone and claystone that lie below it.

The orange slopes above the Capitol Reef bed belong to the Owl Rock member, which caps the Chinle formation in Capitol Reef. The siltstone and claystone of this unit aren't crumbly like the two underlying members, telling us that volcanic activity had ceased by this time. Along with the usual siltstone and mudstone, this unit contains some thin beds of limestone. During this part of late Triassic time, a large lake covered much of what is now the Colorado Plateau. The limestone beds were deposited in periods when the water was clear and calm, allowing carbonate-producing organisms to flourish. The siltstone is evidence of influxes of mud into the lake. The Capitol Reef area was near the edge of this lake, where the water was shallow. Here, mud transported by rivers rained down on the lake bottom much of the time. Carbonate-producing organisms, which prefer warm, clear water, found this a difficult environment in which to live; thus the limestone beds in this area are thin. Farther to the south and east, toward the Four Corners region, the water was deeper and influxes of silt were less common. There, carbonate-producing organisms flourished and the limestone beds are much thicker.

Three members of the Chinle formation and the overlying Wingate sandstone, viewed from stop 2. The resistant sandstone at center is the Capitol Reef bed at the top of the Petrified Forest member.

The marshes, lakes, and streams that deposited the various members of the Chinle formation tell us that the climate during late Triassic time was relatively wet. Near the end of this period, things began to change. At the top of the Owl Rock member, large mud cracks indicate that things were drying out. This shift continued into early Jurassic time. The Wingate sandstone, which forms the imposing orange wall at the top of this outcrop, is further evidence of this transformation.

The Wingate sandstone was deposited in a vast desert that covered most of Utah, northeast Arizona, and bits of Colorado and New Mexico. Large-scale crossbeds in the cliff above you indicate that this was a windblown dune field. You can see these crossbeds in some of the fallen blocks as well. Examine a few of the large blocks of Wingate that litter the ground around you. The grains that make up this sandstone are all about the same size, and nearly all of them are quartz, a very tough mineral. This indicates that these sand grains were blown around for a long time over great distances. Over time, those grains composed of minerals of a lesser constitution were worn away, leaving dunes composed almost exclusively of quartz. The cement that binds the grains contains iron oxide, which weathers to produce the orange color of the rock.

The parking area at stop 3 is near the level of the contact between the Wingate sandstone and the overlying Kayenta formation. Most of what we'll discuss here can be seen right around the parking area, but walking 0.25 mile or so up the Hickman Bridge Trail will be much more

*Block of sandstone at stop 2 displaying two sets
of ripple marks in different directions*

instructive. As you begin the walk, keep your eyes on the wall of rock to your left. Conditions were still arid when the Kayenta was deposited, but a system of rivers flowed west out of the Ancestral Rocky Mountains, bringing an influx of new sediment and also reworking the sands of the Wingate dunes. The Kayenta is predominantly sandstone, but there are also beds and lenses (lens-shaped deposits) of siltstone and conglomerate, rock types not present in the underlying Wingate. If you look closely, you'll not only see some nice siltstone lenses, you'll also see bits of mud that were torn out of them and incorporated into the sandstone. The Kayenta also contains crossbeds, but they're smaller in scale than those in the Wingate. Crossbeds formed by river deposition don't attain the large size of those formed by wind.

About 0.25 mile up the trail, the view opens up to include the overlying Navajo sandstone, the third in this series of magnificent early Jurassic sandstones. The pronounced difference in the weathering patterns of these two units indicates that they were deposited in quite different environments. The tree-covered ledges of the Kayenta contrast sharply with the barren, massive cliffs and domes of the Navajo. Sandstone and siltstone erode at different rates, and a rock unit like the Kayenta, which contains both of them, weathers unevenly, a process known as differential weathering. Such weathering formed the series of benches and ledges

The Kayenta formation (dark rock, foreground) *displays its ledgy nature near the Hickman Bridge trailhead. Behind it, a dome of Navajo sandstone.*

you have been traversing. The Navajo, like the Wingate, is composed almost exclusively of sandstone. Its relatively uniform resistance to erosion results in the walls and domes it's known for.

Arid conditions continued during deposition of the Navajo sandstone. In the last part of early Jurassic time, a huge sea of sand spread slowly down from the north, choking and then burying the Kayenta river system. This dune field covered an even greater area than its predecessor during deposition of the Wingate sandstone and is considered to have been one of the largest in the earth's history. Sets of crossbeds up to 60 feet high attest to the size of its dunes. This sandstone, like the Wingate, is composed almost exclusively of quartz grains. This consistency in composition is one of the main factors influencing the weathering pattern of both of these units. The Wingate sandstone, as we saw at stop 2, forms tall, sheer cliffs. The Navajo sandstone forms sheer cliffs too, but it also forms the domes you see around you. The view of Capitol Dome, the park's namesake, from the parking area at stop 3 offers an easy opportunity to see the contrast in weathering patterns between the ledges of the Kayenta formation and the massive, crossbedded Navajo sandstone. Vignette 8 tells the story of the Navajo sandstone and the immense desert that produced it.

By middle Jurassic time, things began to change once again. The climate was still arid, but due to a combination of subsidence and a rise in sea level, a shallow arm of the sea spilled into a trough west of the Capitol Reef area, this time from the north. The sediments deposited in this region during that time record the complex relationship that developed between sea level, climate, and tectonic activity. Proceed to stop 4, where we'll continue our discussion.

Here at Fremont Falls you can see the erosive power of water in action as it slices a narrow slot into the upper portion of the Navajo sandstone. Look to the northeast across the river, where you'll see an exposure of the uppermost Navajo sandstone with the overlying Page sandstone, a formation of middle Jurassic age. The contact is where the color changes from nearly white to brown. The crossbedded Page sandstone provides evidence that dune fields still existed in this area east of the sea in middle Jurassic time. During this same time, siltstone, mudstone, and limestone were being deposited under shallow marine to tidal flat conditions a bit farther west. The thin, darker band of siltstone within the Page sandstone represents a short period when the sea expanded eastward and tidal flat conditions prevailed here as well. These conditions returned during deposition of the overlying Carmel formation, seen at the top of the outcrop. At that time, the shallow sea spread eastward and layers of mudstone, siltstone, and gypsum smothered the dune field.

At stop 4, on the south side of Utah 24, crossbeds in the Navajo sandstone have been deformed. The dark rock on top is Page sandstone, while the lighter-colored Carmel formation caps the cliff to the right.

At stop 5, the complex relationship between two depositional environments is displayed nicely in the Entrada sandstone. Walk up to the outcrop there and examine it closely. You'll see a series of layers of sandstone and siltstone with thin, crosscutting veins of gypsum. To the south and east of here, the Entrada is predominantly a crossbedded sandstone that was deposited in a large dune field similar to the one that formed the Navajo sandstone. The arches in Arches National Park and the goblins in Goblin Valley are carved from these Entrada dunes (see vignettes 30 and 21). To the north and west, tidal flat conditions prevailed, resulting in shale, siltstone, and gypsum deposits like those we saw in the Moenkopi. This area of Capitol Reef was within the transition zone between these two regimes. The sandstone layers denote times the dune fields dominated the landscape. Look closely and you'll see crossbeds in these layers. While not of the grand scale of the crossbeds in the Navajo sandstone, they still speak of a desolate, windswept landscape. The siltstone layers record times when the sea pushed south and east and tidal-flat conditions predominated. This alternation of rock types indicates that this was the edge of the dune field. If you drive along Notom-Bullfrog Road, you'll

Thin veins of gypsum cut across sandstone and siltstone beds of the Entrada formation.

notice that the sandstone layers thicken as you travel south, toward the interior of the dune field.

From the parking area, you can see the light greenish gray beds of the Curtis formation capping the Entrada sandstone outcrop to the east. Composed of layers of sandstone, siltstone, and limestone, the Curtis formation represents another incursion of the shallow sea from the north. The Curtis formation thins substantially as you travel from north to south in the park, a response to the shallowing of the basin in this direction.

The sea once again receded, leaving a broad tidal flat in arid south-central Utah. The sandstone, reddish brown siltstone, and white gypsum beds of the Summerville formation, which records this change, can be seen at our sixth and final stop, at the eastern entrance of the park. This unit crops out at the base of the cliff across the river to the north. Ripple marks and mud cracks are common in this unit, attesting to repeated episodes of inundation and drying out.

An unconformity of regional extent exists between the Summerville and the overlying Salt Wash member of the late Jurassic Morrison formation. After this period of nondeposition, the climate became wetter and large dinosaurs became common. The Morrison formation is world famous for its dinosaur fossils. The Salt Wash member consists of mudstone, sandstone, and conglomerate. Its thick sandstone beds contrast sharply with the underlying Summerville. When the Salt Wash member was deposited during late Jurassic time, mountain building was occurring to the west in Nevada, and rivers flowing east out of those highlands deposited this variable rock unit.

Look downriver to the east to see the Brushy Basin member, which overlies the Salt Wash member. These rounded, colorful hills, reminiscent of the Monitor Butte member of the Chinle formation at stop 2, are composed of mudstone that was deposited on broad floodplains. This mudstone, like the Monitor Butte claystone, contains clay derived from volcanic ash, which expands when wet, forming the now familiar crumbly surface texture.

Near the end of late Jurassic time, conditions were changing again. To the west, mountain building continued at an increased pace, and this was a time of erosion rather than deposition. Later, during late Cretaceous time, a broad interior sea encroached from the east, initiating another episode of deposition. Our story ends here for now, though we pick it up again about 20 miles south of here in the next vignette, at a sandstone unit that contains an oyster reef. By taking a series of rock units and looking at each individually, we've pieced together a geologic history for this small region of southern Utah. We've seen how a series of deserts, river systems, swamps, and shallow seas left their mark on this area long ago. The Waterpocket Fold aided us in this endeavor by making such an impressive stratigraphic section so accessible. In this same way, geologists have been able to piece together large portions of the earth's history, one fragment at a time.

The thin-bedded Summerville formation is exposed at stop 6. The thick sandstone beds that cap it belong to the Salt Wash member of the Morrison formation.

20 SOMEONE PASS THE TABASCO AND A COLD BEER!
An Oyster Reef in the Desert

Driving south along Notom-Bullfrog Road, one's eyes naturally wander to the west. There, the brilliantly colored rocks of the Waterpocket Fold jut skyward like the immense dorsal fin of a 100-mile-long sailfish. Striking as this structure may be, there are many other somewhat more subtle places of interest to explore in this vast landscape. About 23 miles south of Utah 24, you'll come across one of the more unlikely of these—an ancient oyster reef.

Oysters may seem out of place in the present-day landscape here, but as we learned in vignette 19 on Capitol Reef, times change and so do environments. Follow the directions in "Getting There" to the ridge that contains the oyster reef. The light brown rock of the ridge is part of the Dakota sandstone, a very widespread and consistent rock unit found throughout much of the West. Of late Cretaceous age, the Dakota sandstone is divided into upper and lower units. The lower unit consists of gray to light brown, fine- to medium-grained crossbedded sandstone. It contains lenses of conglomerate, as well as thin-bedded, carbon-rich mudstone and thin coal seams. Fossils are not abundant in the lower Dakota, although some petrified wood has been found.

What's exposed here is the upper unit of the Dakota, which consists almost entirely of fine-grained, thin-bedded quartz sandstone. These sandstone beds tend to form rather continuous ledges. They also tend to be fossiliferous, as you may have noticed. Let's concentrate on the fossil shells that are so plentiful here.

Take a look at some of the specimens that have weathered out of the bedrock. You'll notice that there are a few different types of shells present, but the majority of them belong to the group of mollusks known as pelecypods. Pelecypods are bivalves; that is, their shells consist of two parts, called valves, joined together by a hinge. Modern clams, for example, are bivalves. The largest shells you see scattered here are oysters of the genus *Flemingostrea*. These are the most common fossils in this part of the Dakota, and if you search around a little, you'll find large concentrations of them, known as reefs, in the bedrock. Let's stop for a minute, though, before we proceed. It's easy to have someone say, "These are oysters," and just agree without giving it much thought. But how do we know these are oysters? What is it that makes an oyster an oyster?

What truly makes an oyster an oyster is a deeply philosophical question and may never be answered with complete satisfaction, but if we can look at these shells and understand what makes them different from other pelecypods, we are at least taking a step in that direction. Paleontologists generally don't have the advantage of looking at the internal organs or

---

GETTING THERE

From the east entrance of Capitol Reef National Park, take Utah 24 east 0.1 mile to Notom-Bullfrog Road (see map, p. 146). This turnoff is 39 miles west of Hanksville. Go south for 23.7 miles on Notom-Bullfrog Road, a well-graded dirt road. Watch for washouts due to summer thunderstorms. The ridge of light brown Dakota sandstone containing the oyster reef will be on your left (east). Park on the side of the road and walk about 200 feet to the ridge.

The low ridge of Dakota sandstone to the east of the road

Examples of Flemingostrea. *The two specimens at the top are left valves, and the one at the bottom is a more flattened right valve. Guitar pick for scale.*

other soft parts of an organism to help identify it. In rare cases, the soft parts are preserved, but usually paleontologists must base identifications solely on the hard parts. Pick up one of the larger *Flemingostrea* shells. We've already established that these are pelecypods, but let's check that by making sure you can see two valves. Many of the specimens are broken, but you should be able to find some with both valves intact. As with the common clams and scallops familiar to seafood lovers, in most pelecypods both valves are of similar shape and size. These are called equivalve. In *Flemingostrea*, however, the two valves are of very different shape. One of them is larger and concave; the other is much smaller and flatter. Oysters in general have this inequivalve shape. The left valve is the larger of the two and is usually found cemented to some substrate or to another shell. The right valve is smaller and flatter and fits over the left valve somewhat like a lid.

Now find a nice specimen in which the valves have split apart, revealing the internal shell structure. Hold the left valve with the pointed end, called the beak, up. You may notice some horizontal lines across the beak's triangular surface. This is the hinge area, where the ligament attaches. When the adductor muscle relaxes, the ligament acts sort of like a spring that pops the right valve up, opening the shell at the opposite end. When the adductor muscle contracts, the shell closes. If you have a really nice specimen, you may see a round to semicircular blemish on the interior of the shell just left of center. This is where the adductor muscle attaches. The majority of pelecypods, such as clams and mussels, have two adductor muscles. Oysters as a group have only one. They are termed *monomyarian*, meaning "one muscle."

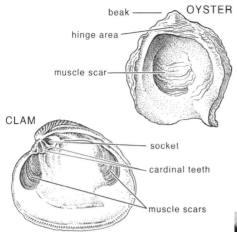

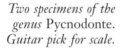

Comparison of the internal structure of the oyster (Flemingostrea) *and the clam* (Mercenaria)*. In paleontology, drawings are often used rather than photographs for showing structure because a photograph represents only one individual, whereas a drawing can represent an ideal form, better showing the characteristics being discussed.*
—artist, Andrew Moore

Two specimens of the genus Pycnodonte. *Guitar pick for scale.*

Another aspect of shell morphology used to classify pelecypods is dentition, or teeth. Many pelecypods have small protrusions, known as teeth, along their hinge line. Cardinal teeth are directly below the beak, and lateral teeth are located at some distance to either side of the beak. Each tooth fits into a corresponding socket in the opposite valve. The teeth help the valves stay in alignment when the shell closes. They also prevent rotational movements from separating the valves. Several terms are used to describe various patterns of dentition. Taxodont, for example, describes a pattern in which many small teeth of variable shape extend nearly the entire length of the hinge line. A heterodont pattern features a toothless space between the cardinal and lateral teeth. If you look along the hinge line of *Flemingostrea*, you will see that there are no visible teeth, a condition described as edentulous. All oysters are edentulous.

Through careful study and comparison of these and other characteristics, paleontologists can determine the various relationships between different organisms. Seeing the differences between *Flemingostrea* and a common clam like *Mercenaria*—inequivalve versus equivalve, one adductor muscle versus two, edentulous versus heterodont—brings us one step closer to knowing what makes an oyster an oyster. If you walk along the top of the Dakota sandstone here, you'll find numerous examples of another type of oyster scattered about. These fossils, in the genus *Pycnodonte*, are often called devil's toenails because of their resemblance to gnarled toenails. Compare these smaller specimens to the *Flemingostrea* shells. Look for the similarities that indicate both these animals were oysters, and the differences that distinguish them.

Now that we're satisfied with our identification of the majority of these organisms as oysters, let's see what we can learn from them. If you pay attention, you'll find that fossils can be great storytellers. Let's see if these oysters can tell us anything about what this part of Utah was like at the time this rock unit was deposited.

One of the guiding concepts in the study of geology is the principle of uniformitarianism. Stated simply, this means that geologic processes such as erosion, deposition, and volcanism worked the same way in the past as they do today. So, for example, observing the processes at work in modern dune fields can help us understand the pattern of deposition that resulted in the formation of a thick, crossbedded sandstone. Applying this idea to paleontology, we can assume that the environmental

This reef features large concentrations of oysters. Lens cap for scale.

requirements of a fossil organism were similar to those of its closely related living relatives.

Modern oysters are generally found cemented firmly to a hard substrate such as a rock or other shells in shallow marine environments. Their relatively thick, flattened shell and the tenacity with which they cement themselves to the substrate are indications that oysters are accustomed to rather high-energy conditions, such as the strong currents and dynamic wave action found close to shorelines. Applying the principle of uniformitarianism, we can postulate that these fossilized organisms also lived in a shallow nearshore marine habitat. Taking this logic one step further, we can also postulate that the upper Dakota sandstone in which these fossils are found was deposited in this same nearshore environment.

Take a look at one of the concentrations of oyster shells in the Dakota sandstone. These reefs begin when a few oysters cement themselves to suitable substrate on the ocean floor. In time, other oysters attach themselves to the shells of those pioneers, and in this way the reef grows. You may note that many of the shells are positioned as they would have been when the oyster was alive, with the larger valve on the bottom. Some of them, however, have been flipped over, indicating that this was indeed a relatively high-energy environment.

This, however, is not the end of the story. The upper Dakota sandstone and the fossils it contains are only one piece of a larger picture. To get an idea of how it all fits together, we must consider the rock units that lie above and below. The rock unit underlying the Dakota sandstone is the Cedar Mountain formation, found only in the northern part of Capitol Reef. Erosion probably removed it from this area. The Cedar Mountain formation has been divided into two units. The lower unit consists of sandstone and conglomerate and was deposited by rivers flowing out of the highlands to the west. The siltstone of the upper unit was probably deposited on broad floodplains as the terrain flattened out and the velocity of the rivers decreased. The lower Dakota, as described above, consists of sandstone, conglomerate, carbon-rich mudstone, and thin coal seams. The conglomerate and coarse sandstone were probably deposited in river channels, and the mudstone and coal seams probably originated in murky swamps. Sediments deposited by wind, rivers, or glaciers or deposited in swamps or lakes are said to be of continental origin, as opposed to marine deposits, which are deposited in the ocean. The Cedar Mountain formation and the lower Dakota are both continental in origin.

Above the Dakota sandstone lies a thick sequence of shale and mudstone that makes up the Tununk shale member of the Mancos shale. Because of the soft nature of the shale, it's easily eroded; therefore, this

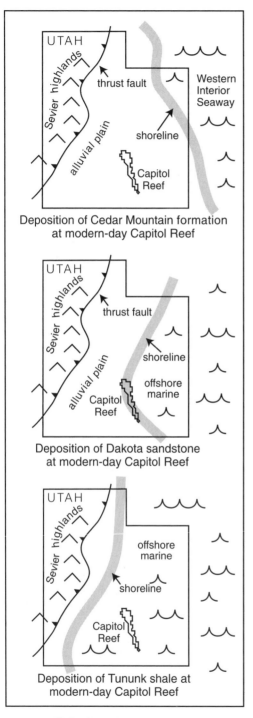

Deposition of Cedar Mountain formation at modern-day Capitol Reef

Deposition of Dakota sandstone at modern-day Capitol Reef

Deposition of Tununk shale at modern-day Capitol Reef

Paleoclimatic reconstruction

unit isn't well exposed here. The Tununk shale underlies the valley just to the east and can be seen at the base of the next sandstone ridge in that direction. Shale, made up of very fine sediment, usually indicates deposition in a deepwater marine environment.

You may notice a pattern developing here, beginning with the continental Cedar Mountain formation and lower Dakota, continuing through the shallow marine upper Dakota, and ending with the deep marine Tununk shale. This pattern is caused by migration of the ocean shoreline due to a rise in sea level. This spread of the sea over land areas during a rise in sea level is called transgression, and the sequence of rocks deposited during this time is referred to as a transgressive sequence. Here's how it worked in this area: The sequence began with deposition of the Cedar Mountain formation by rivers. This deposition was followed by a period of erosion during mid-Cretaceous time that completely stripped away this formation in the southern part of what is now Capitol Reef National Park, although much of it remained farther to the north. Late Cretaceous rivers eventually began to deposit the lower Dakota. These rivers carried large amounts of sediment from the highlands to the west to a vast inland sea located to the east. At its height, this sea submerged much of central North America—from eastern Kansas to Utah and from the Gulf of Mexico into Canada. Deposition of the lower

Dakota began as coarse sand and gravel accumulated in low areas that were cut into the bedrock during the episode of erosion. Farther east, where the rivers interfaced with the sea, the current slowed considerably and the rivers dropped much of their sediment load. This created marshy conditions, and plants and other organic material accumulated to form the mudstone and thin coal seams present in the lower Dakota. As sea level began to rise and the shoreline migrated west, so too did this area of interface, leading to deposition of the fossiliferous shallow marine sandstone of the upper Dakota formation on top of the carbon-rich mud. You're standing in this part of the sequence.

As more time passed, the sea in this area deepened and sand was deposited along the new shoreline to the west. Only extremely fine-grained sediments were carried out this far from shore. These fine particles drifted down to the seafloor to form the thick layers of mud that later became the Tununk shale. The drab gray shale that makes up this unit underlies the low area to the east.

At some point the sea stopped rising, a condition known as sea level highstand, and the shoreline ceased migrating west. This happened sometime during the deposition of the marine shale of the Tununk. Following this highstand, the process began to reverse. Worldwide lowering of sea level, changing tectonics in the basin, and possibly other factors caused

View east from near the oyster reef. The light-colored, tree-studded ridge in the foreground is the Dakota sandstone. The Tununk shale underlies the low area behind it. A low ridge of Ferron sandstone can be seen at center left. The tall cliff in the background is composed of Bluegate shale capped by sandstone of the Muley Canyon member.

the shoreline to recede back toward the east. Here in Capitol Reef, you can see evidence of this shift in the next sandstone ridge to the east. That ridge is composed of the Ferron sandstone member of the Mancos shale. Just as the Dakota sandstone you're standing on records the time when the westward-migrating shoreline passed through this spot, the lower portion of the Ferron sandstone is a record of the shoreline's retreat back to the east.

The sequence of rocks deposited as sea level lowers is called a regressive sequence, and it's a mirror image of the transgressive sequence. Progressively sandier nearshore sediments overlie the deepwater shale and mudstone. These sediments grade into the coarser-grained sandstone that was deposited at the shoreline interface; together they make up the lower part of the Ferron. This particular cycle of transgression and regression, called the Greenhorn Cycle, is the first of four major cycles that occurred in late Cretaceous time. After deposition of the lower part of the Ferron, an interval of erosion took place. Later, sea level began to rise once again and the sandstone of the upper part of the Ferron was deposited over this erosion surface. This began the second major transgessive-regressive sequence.

Off in the distance to the east, you can see evidence of these repeated cycles. The lowlands beyond the Ferron are underlain by the Bluegate shale which, like the Tununk shale, was deposited in deep marine conditions. This unit also makes up the lower part of the cliff beyond. That cliff is capped by the Muley Canyon member, another resistant sandstone unit deposited as the receding shoreline passed by on its next journey east. And so the cycles continued for the rest of Cretaceous time. It's rather amazing that a pile of fossils can take you on such a journey. But by figuring out what they are and what they can tell us, these oysters in the desert take us a long way toward understanding part of the geologic history of Capitol Reef.

21 A WEATHERED ARMY ON THE MARCH
Goblin Valley State Park

As you travel across the Colorado Plateau, uniquely shaped and brightly colored geologic formations engage your imagination and tantalize your senses. The processes of erosion and weathering sculpted these formations, and in Goblin Valley State Park, they've created some of the most distinctive hoodoos you'll see anywhere, especially in such profusion. A hoodoo is a column or pillar of rock produced by erosion or weathering,

A camel and a turtle rest atop sandstone pedestals in the setting summer sun.

often with an eccentric form. In Goblin Valley, the hoodoos are known as mushroom rocks due to their rounded shapes atop pedestal foundations. But the shapes of these rocks—sometimes human, sometimes more fantastic—have earned these formations another name as well: goblins.

The native people who traveled through this area and left pictographs in the nearby San Rafael Swell must have known of this strange valley. During the 1800s, cowboys searching for stray cattle were the first recorded Euro-Americans to come across it. In the 1920s, Arthur Chaffin,

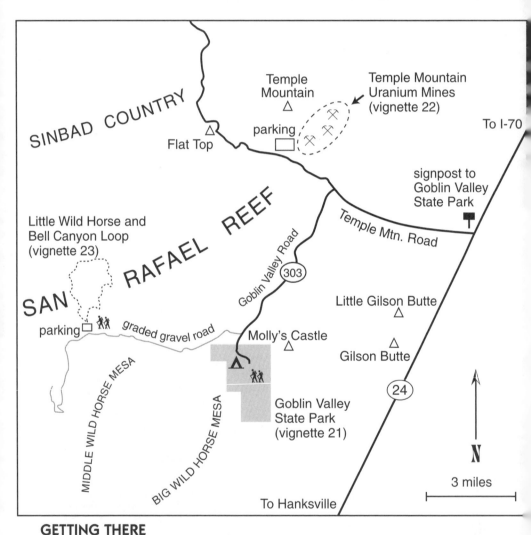

GETTING THERE

From the town of Green River, go 12 miles west on I-70 to exit 147, then take Utah 24 south for 26 miles to Temple Mountain Road. Alternately, take Utah 24 north from Hanksville for 21 miles to that same junction. Turn west at the intersection—there's a sign for Goblin Valley State Park—and proceed 5 miles to Goblin Valley Road (Utah 303); turn south (left) and go another 6 miles to the park. The valley itself is less than 1 mile from the campground.

The view east across Goblin Valley from the overlook

owner of the ferry across the Colorado River at Hite, stumbled upon the valley as he searched for a route from Caineville to Green River. Returning in 1949, he explored and photographed it, bringing greater public awareness to the valley. Chaffin called it Mushroom Valley. In 1964, Utah designated the valley and a total of 3,654 acres as Goblin Valley State Park.

Let's start at the park's overlook. The valley is about 0.5 mile wide and 2 miles long. Across from the overlook, a cliff that trends north-northeast defines the eastern side of the valley. Five sediment deposits, four from Jurassic time and one from Quaternary time, are visible in the park. The cliff is formed of two Jurassic deposits: the reddish brown, goblin-forming Entrada sandstone and, above it, the lighter-colored Curtis formation. From the overlook, you can see how the goblins march forward out of the sandstone, especially from the cliff base as they travel out onto the valley floor. Erosion and weathering lead them into existence.

Erosion is the process by which a moving medium such as wind or water excavates and transports rock material. Weathering, on the other hand, is the physical disintegration of rock by such agents as ice, rain, snow, or temperature changes. Weathering can also occur by chemical decomposition, in which air or water chemically alters the rock, causing

Even the smallest goblins stand high above the valley floor. The resistant and nonresistant layers are easily distinguished.

its decay. So by definition, weathering doesn't include transportation of material, though weathered material is more likely to erode. Erosion and weathering act in concert to isolate, sculpt, and wash away debris. The metamorphosis of the goblins before you starts on the cliff wall. Fault systems traversing the valley have created small fractures, or joints, up to 1 inch long in the Entrada sandstone. Some of these joints intersect, forming corners and edges. With more surface area exposed to weathering, corners and edges break down more easily than an unfractured rock face. Through continued erosion and weathering, the fractures widen, separating standing pieces of rock from the wall to form hoodoos and rounding them in a process known as spheroidal weathering.

Interbedded layers of shale and siltstone in the Entrada formation erode and weather more quickly than the harder sandstone. These varying degrees of resistance result in ledges of sandstone that overhang indented, softer layers of shale and siltstone. The sandstone cap protects the softer layers, granting a hoodoo some measure of longevity, but eventually it too must give way (hoodoos are also discussed in vignette 17). After a hoodoo's cap topples, erosion and weathering rapidly reduce the shale and silt layers to poorly defined mounds on the valley floor. Over time, the toppled sandstone cap succumbs to the same fate, though much more slowly. If you haven't already, descend into the valley and

As weathering proceeds, goblins march out of this wall of Entrada sandstone.

wander among the goblins. See if you can find any in these later stages of development.

Geologists speculate that another factor may contribute to the unusual shapes in Goblin Valley. Variations in the amount of mineral cementation between grains of sand within the rock create varying resistance to erosion and weathering. The end results of this and all the factors already discussed are Goblin Valley's peculiar yet familiar and evocative shapes. The best time to view and photograph them is an hour or two before sunset on a clear day. In the soft light and deep shadows of evening, watch the forms of animals, caricatures of people, and other fantastic shapes emerge vividly from the rock.

During middle Jurassic time, eastern Utah, including Goblin Valley, was a basin created by uplift of highlands to the west and the Ancestral Rocky Mountains to the east. Ocean waters periodically flowed down from Canada and filled the basin with an inland sea. The siltstone, shale, and sandstone from which Goblin Valley is carved have their origins in sediments eroded down from the surrounding highlands and deposited in this sea, where they settled in tidal mudflats and interbedded with the Entrada sandstone. Ripple marks in the Curtis formation in the park record fluctuations in sea level and wave activity on that ancient ocean's beaches. In later Jurassic time, the sea fully retreated. River systems replaced it and continued deposition. Ultimately erosion, spurred by the Colorado Plateau's uplift 10 million years ago, excavated the valley and, together with weathering, carved the bizarre tableau before you.

When you first approached the park entrance, Wild Horse Butte (elevation 5,760 feet) was on your right. All four Jurassic sediment deposits visible in the park—the Entrada, Curtis, Summerville, and Morrison formations, from oldest to youngest—appear in the butte and can be seen from below. At the bottom, you may recognize the Entrada sandstone, tinted red by iron oxide cement, and the Curtis formation. The greenish gray color of the Curtis formation comes from glauconite, an iron-rich clay mineral associated with ocean sedimentation. The Summerville formation, the chocolate- to bone-colored unit second from the top, formed in a tidal flat after the inland sea retreated. Numerous white gypsum veins in the Summerville probably resulted from periodic evaporation in these tidal flats or in ponded seawater. The Morrison formation, the youngest layer, is visible only at the very top of Wild Horse Butte. This formation developed as the river systems that dominated late Jurassic time here deposited a conglomerate of silt, clay, sand, and gravel on its

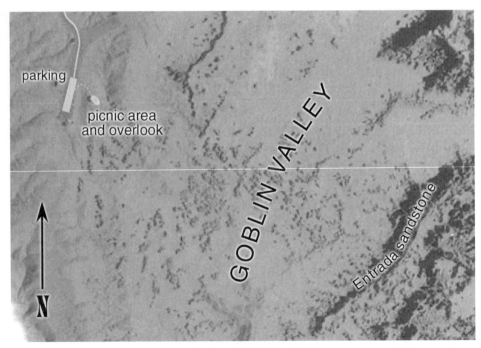

Aerial photo of Goblin Valley

floodplain. More recent Quaternary deposits—talus, alluvium, and wind dunes—are seen on the plains on either side of the road as you approach the ranger station on Goblin Valley Road.

Goblin Valley also offers glimpses of the larger geologic context, with views of the Henry Mountains to the south and the San Rafael Reef to the west. The Henry Mountains, an intrusive igneous rock formation similar in structure to the La Sal Mountains (discussed in vignette 32), were one of the last ranges in the continental United States to be named and surveyed. You can also see them from the 2- to 3-mile Curtis Bench Loop Trail that begins in the Goblin Valley campground area. The San Rafael Reef, discussed in vignette 23, is the eastern portion of the San Rafael Swell anticline, an uplift with sloping sides. At the nearby Temple Mountain mining site (described in vignette 22), approximately 7.5 miles from the entrance to Goblin Valley State Park, you can view the remnants of decades of uranium mining operations.

Erosion and weathering have made Goblin Valley a unique geologic, artistic, and meditative experience. As you wander in this sculpture garden, contemplate the expanses of geologic time and the amazingly varied conditions that led to the formation of these distinctive hoodoos.

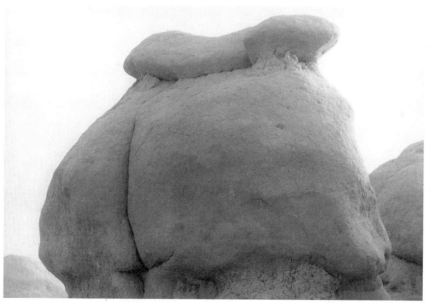

*Some goblins have weathered into forms that are
strikingly similar to anatomical features.*

GETTING THERE

From the town of Green River, go 12 miles west on I-70 to exit 147, then take Utah 24 south for 26 miles to Temple Mountain Road (see map, p. 168). Alternately, take Utah 24 north from Hanksville for 21 miles to that same junction. Turn west and proceed 6.5 miles to the Temple Mountain mine site. You'll drive into a canyon that slices through the San Rafael Reef; keep an eye out for pictographs on the north canyon wall. A Bureau of Land Management (BLM) group campsite marks the location of the mining area, replete with mine entrances, load outs, tailings, dumps, and more. The area also contains several structures, including a few stone houses and buildings, as well as rusted vehicles and other remnants of the village that stood here in the early 1900s and again in the 1950s.

Caution: Wandering around any mine area has potential hazards (hidden vertical shafts, rockfall, and unexploded dynamite), but uranium poses special concerns. The BLM has evaluated these hazards in the Temple Mountain area and posted guidelines. In some posted areas, visitors should avoid continuous exposure beyond 24 hours. In other areas, such as the group camping area, there's no worry; radiation levels are normal for a desert environment. In sum, follow posted guidelines and enjoy the area.

Temple Mountain sits just west of the Waterpocket Fold.

22 RADIATION FEVER
Temple Mountain Uranium Mines

The Temple Mountain mining site is situated in the southeast portion of the San Rafael Swell, a domed uplift of over 2,000 square miles featuring buttes, pinnacles, slot canyons, and more. As seen in vignette 23, some geologic processes within the San Rafael Swell produced stunning and obvious features—for example, the San Rafael Reef. The results of other processes here are less apparent, lying far underground. One of these is uranium, with its invisible radioactive properties and worldwide repercussions. Uranium mining has had an interesting history in Utah, with two boom-and-bust periods. During the mid-1950s, there were eight hundred active uranium mines on the Colorado Plateau alone. The Temple Mountain Mining District, 15,236 acres near Goblin Valley State Park (see vignette 21), is a good place to get a feel for this history, especially since the remains of mining operations here are more accessible than in many places.

Silvery white uranium was discovered in 1789 and named for the planet Uranus, which had been discovered eight years earlier. However, uranium was not isolated as a pure metal from ore until 1841, and its radioactive properties remained unknown until 1896. Uranium is as common in the earth's crust as tin, arsenic, and tungsten. In fact, all rocks contain minute quantities of it, and it's even found in seawater. Every day, we eat a few micrograms of uranium in our food. Uranium is all around us. It occurs in higher concentrations in primary ores, such as uraninite and pitchblende, and in lower concentrations in secondary ores, such as carnotite. Primary ores form when metals are first introduced into rock and mineralized. Secondary ores form during later stages.

175

Most mine shafts have been gated or, like this one, cemented shut. There are serious hazards associated with poking around old uranium mines, including hidden vertical shafts, rockfall, unexploded dynamite, and radon gas.

Nearly 90 percent of uranium production from secondary ore deposits is from carnotite. Secondary ores are usually brightly colored—yellow and green, for example—from an oxidation process, while primary ores are duller black and gray.

While working in a Colorado gold mine, a Cornish miner named Richard Pearce recognized uranium in pitchblende, the dull black substance that miners were discarding. Subsequently Pearce dug the nation's first uranium mine in Gilpin County, Colorado, in 1871. The first recorded uranium mining on the Colorado Plateau got started in 1881, though the Ute and Navajo, who used brightly colored carnotite in their war paint, probably surface mined it prior to that time. In the 1880s and 1890s, uranium was used in glass and pottery manufacturing, experimentally in photography, and to color dye and ink. The first of the two uranium booms in Utah occurred after a French engineer found uranium in carnotite on the plateau in 1898. Carnotite, which occurs primarily in sandstone and is abundant on the Colorado Plateau, demonstrates an interesting characteristic of uranium: an affinity for organic material such as carbonized wood fossils and asphaltic material. This feature is key in the Temple Mountain ore deposits, as we'll explain later.

A miner's cabin remains as a testament to uranium fever.

That same year, 1898, Pierre and Marie Curie discovered the "miracle element" radium, one of several elements uranium produces as it decays. They found that all uranium ores contain trace amounts of radium. Beliefs in radium's curative and medical powers drove the early twentieth-century uranium mining boom. The medical community experimented with it in cures for diseases, including cancer. Entrepreneurs touted the benefits of radioactive baths, injections, salves, and inhalers. Radium was also used for illumination in compasses, watch faces, airplane dials, and the like. Only much later was it discovered that radium is more radioactive than uranium and capable of producing radon gas, also a health hazard. Exposure to radium and to uranium ores damaged the health of many miners and researchers, including the Curies. Many uranium miners succumbed to lung cancer before safe levels of exposure to uranium and other radioactive elements were established. Sadly, that wasn't until well after Manhattan Project scientists warned of possible hazards.

The first claims at Temple Mountain were staked in 1898. Exploration of the area and premining development work lasted until 1914, when mining for uranium and vanadium began, continuing into the 1920s. Vanadium, a metal that improves the tensile strength of steel, was important in steel production during World War I. Temple Mountain was also a significant source of radium, found in carnotite and leftover mine tailings, until discoveries elsewhere in the world drove prices down. Mining waned in this area from the late 1920s until the late 1940s before resuming at a fairly high level through the 1960s. From the late 1940s through the mid-1950s, the district produced 261,000 tons of uranium ore, with the largest ore body yielding 20,000 tons. This second boom was prompted by the cold war and development of nuclear energy technologies, which ushered in the atomic age. To illustrate uranium's capacity in electrical power production, it would take only 24 tons of enriched uranium to support a million-kilowatt power station for a year after startup; the same task would require over 3 *million* tons of coal. Uranium mined around Temple Mountain was used in post–World War II Manhattan Project initiatives and later atomic bomb development, as well as for other purposes.

You may notice that the mines are primarily centered on one stratigraphic layer, a layer containing a material that looks like black asphalt. In fact, this is asphaltite, a carbon-rich material that, here at Temple Mountain, hosts uranium and vanadium. Most research leans toward asphaltite being a petroleum derivative, thus originally organic. The asphaltite here formed when asphaltic material saturated uraninite- and carnotite-bearing sandstone. The saturation is so great here that the Temple Mountain uranium deposits have been accorded a unique name: uraniferous asphaltite deposits. These deposits primarily occur in the Moss Back member of the Chinle formation, with lesser amounts in the Moenkopi formation and Wingate sandstone.

But how did the uranium-rich asphaltite form? The most widely accepted theory holds that it formed through millions of years of sediment deposition, regional deformation, and eventual hydrothermal alteration of existing petroleum deposits by magmatic fluids followed by uranium emplacement. According to this theory, carbonate rocks in formations beneath the Moss Back sandstone and Wingate formation dissolved, causing Moss Back and Wingate units to drop hundreds of feet and creating fractures and faults. Extremely hot uranium-bearing hydrothermal solutions migrated from local magma bodies through the faults and fractures. These solutions interacted with existing petroleum in porous sediments of the collapsed structures, producing the uraniferous deposits. Bleached rocks within the collapse structures and evidence of past and present hot spring and igneous activity in the area are among the clues supporting this hydrothermal theory.

As you stroll around the Temple Mountain mining area, evidence of past mining activities are all around you. You'll notice rounded mine

Mine shafts abound in this horizon.

shaft openings in the rock walls, some over 6 feet in diameter. The shafts are now sealed with metal grates and concrete. Dirt roads in the area lead to mines, headworks, and piles of tailings, waste rock that was dumped down hillsides during mine excavation. Some rock walls still bear concussion scars from dynamite.

The present abandoned state of the Temple Mountain mines is a reminder of the larger political story uranium played a part in during the twentieth century. Among other things, this is a story of deep ambivalence. After proving the feasibility of a nuclear chain reaction, scientist Leo Szilard, who with Einstein recommended a nuclear program to President Roosevelt in 1939, stated, "The world is headed for trouble. The world is headed for grief." Similarly, upon witnessing the Trinity test, the first nuclear explosion, in July 1945, "father of the A-bomb" J. Robert Oppenheimer quoted the *Bhagavad-Gita*: "I am become Death, the shatterer of worlds." The uranium mined here fueled these devastating developments and contributed to an immense stockpile of weapons that has cost the United States over $5 trillion from 1940 to 1996. While prospects of wealth drive most mining stories, power struggles, political posturing, and now terrorism are driving factors in the story of uranium.

Now the United States is considering dismantling its more than ten thousand nuclear weapons and, worldwide, trying to manage the availability of enriched uranium. But uranium and other radioactive elements do not go gentle into that good night. Depending on their structure, these elements will take hundreds of thousands if not millions of years to decay. How to safely manage spent radioactive materials and the various stockpiles of uranium worldwide will remain an issue for years to come—one that also involves geology, as the merits of various underground storage facilities are considered. But that's another tale.

Mine tailings at Temple Mountain

23 SLITHERING THROUGH A SLOT CANYON
The San Rafael Reef

In southern Utah, a land of tens of thousands of canyons, slot canyons are still something special. Though often as deep as other canyons, they are much narrower—sometimes barely wide enough to allow passage on the channel floor. From within them, the sky is usually visible as only a narrow ribbon of blue, but the shade these canyons offer is welcome on a hot summer day. Smooth, sculpted rock, occasional pools, waterfalls, and the play of sunlight against sinuous walls make slot canyons exceptionally photogenic. There is, however, a downside to slot canyons. During storms they can fill rapidly with water to depths of 20, 30, even 50 feet, and their architecture doesn't allow easy escape. Unwary travelers have been trapped and even killed by rising water in these confined areas.

This vignette takes you on a hike through a slot canyon into the heart of an uplifted mass of rock called the San Rafael Swell. Little Wild Horse Canyon is a slot canyon that cuts from west to east through the San Rafael Reef, a prominent sandstone ridge at the eastern edge of the San Rafael Swell. The word *reef* may seem odd here in the desert, but many early settlers were led west by former seafarers who used the word *reef* for any obstruction to travel, whether on sea or land. On this hike, you'll travel through Little Wild Horse Canyon to a jeep trail on the other side of the reef, south on that road, then return via Bell Canyon to the parking area.

As we said in "Getting There," the difficulty of this trail can vary greatly depending on moisture conditions. We've been lucky enough to visit Little Wild Horse and Bell Canyons in both dry and wet conditions. We say lucky because, like most of the Colorado Plateau, the San Rafael

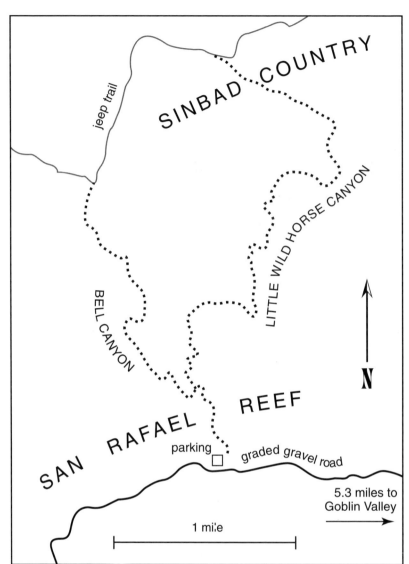

Little Wild Horse
and Bell Canyon
Loop Trail

GETTING THERE

From the town of Green River, go 12 miles west on I-70 to exit 147, then take Utah 24 south for 26 miles to Temple Mountain Road (see map, p. 168). Alternately, take Utah 24 north from Hanksville for 21 miles to that same junction. Turn west at the intersection and proceed 5 miles to Goblin Valley Road (Utah 303). Drive south toward Goblin Valley; after 6 miles you'll reach a crossroad just beneath Wild Horse Butte. Turn right (west) onto a graded gravel road and go 5.3 miles to the marked parking area at the mouth of Little Wild Horse Canyon. The entire Little Wild Horse and Bell Canyon Loop Trail is 8 miles long and takes 4 to 5 hours to hike (that time can double if conditions are wet). The trail is usually described as relatively easy, but some agility may be required. There's an information board, trail register, and toilet at the trailhead, but water is not available.

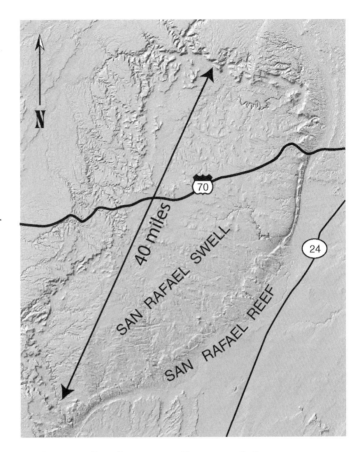

San Rafael Swell and Reef. Arrow shows length of swell's central uplifted portion.

Reef is arid, so it's unusual to see abundant water. But water is important here, particularly for our purposes, as it has carved this canyon. We first hiked the loop trail the day after heavy rains in this part of southern Utah (after the flood danger had passed). The sky was clear, but the canyon was full of silty pools and thick mud. We spent most of the day either wading through pools or climbing around the ones that were too deep to wade. The presence of water made it an interesting hike but doubled the amount of time required.

From the parking area, walk up Little Wild Horse Creek, an open wash shaded by old cottonwood trees. A small waterfall along this initial stretch can be bypassed to the left. Water rushing over these falls during storms has created the depression at the base, known as a plunge pool. About 0.5 mile from the trailhead, a tributary enters the main channel from the right. Bell Canyon continues straight ahead; you want to continue up Little Wild Horse Canyon, the tributary to your right. Be alert, as it's easy to miss this intersection. Sometimes a rock cairn or arrow shows the way, but floodwater can destroy these markers.

Extended lengths of extremely narrow sections of Little Wild Horse Canyon begin less than 0.25 mile upstream from this intersection. You'll discover sections so tight that you have to turn sideways and slither through them. These stretches are broken up by short open valleys where tributary canyons join Little Wild Horse Canyon. A ubiquitous characteristic of slot canyons, and one that contributes greatly to their aesthetic quality, is hydraulically sculpted bedrock. As water rushes through these channels during storms, its load of sand and silt scours the rock, producing streamlined knobs and flutes, which are long, smooth grooves in the flow direction. Along the lower canyon walls you'll discover weathered rock honeycombs that have been polished into sensuous works of art. You'll maneuver over chock stones, rocks that have fallen from above and become trapped, suspended in the tight passageways above the canyon floor. Look for flood debris overhead, indicating water levels during past storms.

Potholes in the canyon floor retain water long after precipitation events.

As you follow Little Wild Horse Canyon, you're passing through the San Rafael Reef and into the San Rafael Swell, a kidney-shaped anticline approximately 84 miles long and 35 miles wide. (An anticline is a domed upwarp; for more about anticlines, see vignette 6.) The San Rafael Swell rises nearly 2,000 feet above the surrounding land. Older geologic formations exposed at its center are ringed by ridges of resistant younger rock. Imagine placing five or six popsicle sticks on top of one another and orienting them east to west. Push the long ends together so they bend up in the middle until the pressure breaks the upper two or three. The underlying sticks, which represent older rocks, would be exposed beneath the broken layers. Likewise, here in the swell, the rock you see used to be at great depth, where it experienced heat and extreme pressure, both of which caused it to bend without breaking—like the lower popsicle sticks. Millions of years of erosion eventually exposed it at the surface.

Slithering through a slot canyon

In the case of the San Rafael Swell, the horizontal pressure came from a plate tectonic collision between the North American Plate and the Farallon Plate, the Pacific Plate's predecessor. Though the plate collision occurred a considerable distance away—at the present-day California coast—compression caused intercontinental buckling. Approximately 60 million years ago, in early Tertiary time, this buckling created the San Rafael Swell, the Rocky Mountains, and other geologic features, such as the Circle Cliffs in south-central Utah. An orogeny is a period of mountain building, and this particular event is known as the Laramide Orogeny. The Laramide Orogeny occurred after the Sevier Orogeny, an earlier episode of mountain building, and generally affected areas east of the Wasatch Mountains. Erosion subsequently stripped off about 4,000 feet of younger rock from later Cretaceous and Tertiary time, exposing an older sequence of Permian, Triassic, Jurassic, and early Cretaceous rock. Approximately 4,600 feet of stratigraphy is exposed within the San Rafael Swell.

The San Rafael Reef, really the eastern limb of the San Rafael Swell, is a monocline, or one-limbed fold, with a dramatic, near-vertical slope. Sedimentary beds east of the reef are horizontal. At the reef itself, they suddenly bend up at angles that range from 20 to 85 degrees, then flatten out again in the interior of the swell. The west side of the swell features more gradual slopes down to the west. The reef is a massive structure that extends from the Henry Mountains in the south to north of I-70. Over thirty canyons cross the southern San Rafael Reef, and its near-vertical slope and deep slot canyons offer great opportunities for exploration.

Resistant sandstone beds that dip steeply to the east mark the San Rafael Reef.

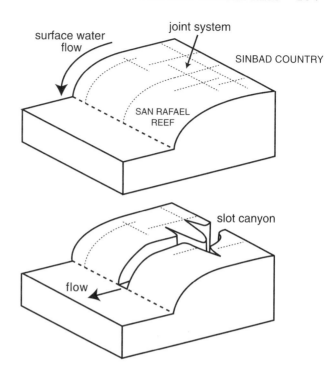

Joints in the bedrock determine the course of future slot canyons.

Within Little Wild Horse Canyon, you see the reef from the inside. As you progress from east to west, notice that the angle of the east-dipping sandstone beds decreases until eventually the beds are horizontal. At that point, you're west of the monocline and in the interior of the anticline. All slot canyons in this area traverse the reef in an east-west direction, slicing cleanly through Navajo, Kayenta, and Wingate sandstones from Mesozoic time. These well-cemented sandstones, resistant to weathering and erosion, can sustain steep cliffs. These units are crisscrossed by systems of fractures called joints, which formed when the rocks here folded into the San Rafael Swell.

The swell's canyons formed as rain that fell in the western highlands flowed to the lowlands in the east. Joints, which indicate planes of weakness, allowed water to move downward through the rock, so channels deepened much more rapidly than they widened. Water is very good at eroding rock, doing so in two ways: hydraulic lifting and abrasion. Hydraulic lifting, a function of high and low pressure caused by turbulence, allows water to pick up loose material from the channel banks and bottom. Abrasion by moving sand and silt simultaneously smoothes and polishes the bedrock over which the stream flows. Here in the swell, erosion is concentrated into the short periods of time in which water actually flows in these canyons.

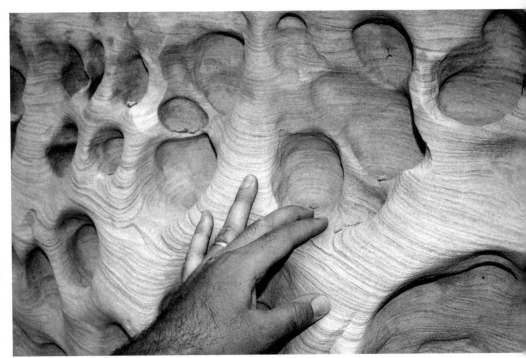

These weathered pits have been smoothed by rapidly moving water and sediment.

Little Wild Horse Canyon opens up on the western side of the reef. Before you is the central area of the San Rafael Swell, referred to as Sinbad Country—another seafaring name. Occasional buttes, spires, and ridges of the Moenkopi and Chinle formations, of Triassic age, rise from a flat and barren plain that's approximately 40 miles long from north to south and 10 miles wide from east to west. Older Coconino sandstone and Kaibab limestone, of Permian age, are also visible. If you want to explore Sinbad Country later, Temple Mountain Road is a good choice. As always, check weather and road conditions first. Along this road you pass Flat Top (a butte), get great views of Temple Mountain, and see some of the spires and buttes that form the San Rafael Swell. You may choose to stop by the Temple Mountain uranium mines, described in detail in vignette 22.

To continue your hike, look for the jeep trail that parallels the back of the reef. Turn left (south) and hike this road over a rise and then east into the top of Bell Canyon. The hike down Bell Canyon is less impressive than that up Little Wild Horse Canyon. The trail is shorter, and since the canyon walls are set farther apart, you don't encounter the breathtaking narrows found in Little Wild Horse Canyon. Still, Bell Canyon has its own charms—brightly colored rock, waterfalls, and sculpted forms—and traversing it gives you another opportunity to observe the orientation of rock layers in the San Rafael Reef. At the top of the canyon, rock units are nearly horizontal. As you proceed eastward down the canyon, the east-facing dip of the rock becomes increasingly steep until it reaches its maximum angle near the front of the reef.

Safety is paramount whenever you explore a slot canyon. In August 1992, ten visitors entered Little Wild Horse Canyon in the morning. Where the canyon opens up on the west side of the reef, they saw a huge storm brewing. Rain and hail assaulted the hikers, and rapidly rising water eventually trapped them. They watched as higher and higher waves surged by, some carrying entire trees. Raging water swept huge boulders downstream and shook the canyon walls. Three hours after the storm ended, water levels dropped to the point where hiking and swimming out through the frigid pools left behind by the storm was possible—though certainly not comfortable. The unhappy explorers finally reached the mouth of the canyon and the parking area, only to discover that the rushing water had treated their vehicles just as it had the trees and rocks. Four-wheel-drive trucks had been carried as much as 0.5 mile downstream, turned upside down, and filled with sand and silt.

This may seem like a sad tale, but it really isn't. All of the hikers survived, which is not always the case with slot canyon floods. In 1997, eleven hikers were killed in Antelope Canyon, a slot canyon in northern Arizona. A storm 5 miles upstream resulted in a sudden flood similar to the one described in Little Wild Horse Canyon. Surging water ripped all the clothes off the only survivor, the expedition's guide. The bodies of some of the dead were pulled from Lake Powell, 4 miles from the canyon's mouth. The moral of this story? Never, ever enter a slot canyon if there is any possibility of rain. In the narrows, there is no escape.

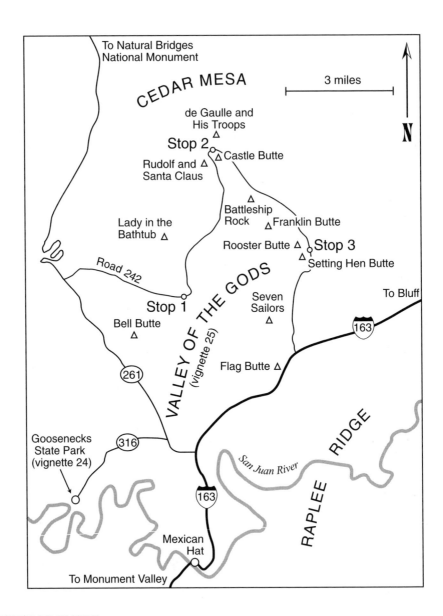

GETTING THERE

From the Utah-Arizona border, drive 25 miles northeast on U.S. 163 through the town of Mexican Hat, Utah, to Utah 261. Drive northwest on Utah 261 about 1 mile to the signpost for Goosenecks State Park at Utah 316. Turn left (west) on Utah 316 and go 4 miles to the Goosenecks Overlook.

24 UPLIFT AND EROSION ON THE COLORADO PLATEAU
Goosenecks of the San Juan River

Goosenecks State Park is very small, essentially just a parking lot and a viewpoint, but the panorama is expansive in terms of both space and geologic time. At the overlook, you stand on the edge of a deep canyon, looking down at the San Juan River 1,100 feet below. Geologists loosely apply the term *gooseneck* to sinuous canyons that resemble the curves of a goose's neck; here, the term refers to the deeply entrenched river to the southwest. To your left, the San Juan River flows downstream from its headwaters high in the San Juan Mountains of southern Colorado. To the right, the river continues downstream to Lake Powell, where it joins the Colorado River, which eventually spills into the Gulf of California. Underfoot, and along the entire canyon rim here, you see a pale gray rock that is quite different from the red sandstones found in Monument Valley to the southwest and Valley of the Gods to the northeast. This gray rock is limestone, made from the remains of marine organisms. Although Goosenecks State Park sits at an elevation of 4,800 feet above sea level, the rock you're standing on formed beneath an ocean hundreds of millions of years ago. Things have obviously changed a great deal since that time.

Exposed bedrock within the San Juan River canyon is Paleozoic in age, more than 245 million years old. There are two formations visible from the overlook: the underlying Paradox formation (middle Pennsylvanian) and the overlying Honaker Trail formation (late Pennsylvanian), upon which you now stand. From here, they look pretty much the same, and both units have their origins in marine environments. But fundamental differences allow geologists to distinguish the two formations.

191

Panoramic view (above and right) *of the Goosenecks of the San Juan River.*
Deeply incised meanders indicate regional uplift and stream erosion.

The Paradox formation was deposited within a marine basin that covered about 17,000 square miles in late Paleozoic time. Cycles of inundation and dessication produced an alternating sequence of limestone, which forms in oceans, and evaporites such as salt, which crystallize as large bodies of water dry up. Reeflike deposits within some of the limestone layers indicate ancient ocean-shelf environments similar to areas where present-day active reefs exist, for example, off coastal Australia and near Caribbean islands. Reefs are agglomerations of the shells of living and dead organisms cemented into a solid mass in a shallow, warm marine environment; they represent a group effort by many different organisms. The Honaker Trail formation, on the other hand, contains cycles of limestone alternating with sandstone, siltstone, and shale. These alternating layers reflect water level fluctuations, regional uplift, and eventual transition from a marine to a terrestrial setting.

The Goosenecks sit in the central Colorado Plateau, a physiographic province characterized by regional uplift during Cenozoic time. The timing of the stages of uplift is still under debate, but most scientists agree that upward movement began with the Laramide Orogeny in early Cenozoic time. (The word *orogeny* refers to a mountain-building event.) The Laramide Orogeny resulted from regional east-west compression associated with a convergent plate tectonic boundary to the west, wherein two plates pushed against each other, producing stresses that folded and fractured bedrock on a very large scale. The Laramide Orogeny formed the massive Rocky Mountains to the east. While the broad Colorado

Plateau contains relatively undisturbed horizontal rock units, it also displays some evidence of Laramide compression.

Compression produces upward folds called anticlines, downward folds called synclines, or a sequence of both anticlines and synclines, all of which are perpendicular to the stress direction. To get a better feel for the relationship between the orientations of stress and folding, push the two ends of a sheet of paper together so that it folds up or down. The fold, whether up or down, is oriented perpendicular to the direction in which you pushed. Goosenecks State Park sits on a north-south trending anticline, an upward fold called the Monument upwarp. Other such features are scattered across much of southern Utah. Some researchers believe the lithosphere beneath the Colorado Plateau became more buoyant due to incorporation of material from the west or because of the behavior of the subducted plate, which was forced beneath the overriding plate at the convergent margin. In either case, plant fossils demonstrate that the Colorado Plateau was as high at the end of the Laramide Orogeny as it is today, and possibly even higher.

Some geologists think that early Cenozoic uplift was followed by a period of epeirogeny, upward movement without internal deformation. Others argue that epeirogeny is unnecessary to explain later uplift, which can be attributed to isostasy. In the introduction, we described general plate tectonic interactions and described the floating balance between the lithosphere and the asthenosphere. Adjustment of that floating balance due to addition of mass (causing lithospheric sinking) or removal of mass (allowing lithospheric rising) is called isostasy. Removal of mass from the Colorado Plateau has occurred by exhumation, the large-scale removal

*The Goosenecks have been cut into the uplifted Honaker Trail formation.
Limestone at the canyon rim formed on the seafloor in a Paleozoic ocean.*

of rock by stream erosion. Sedimentary deposits on the plateau show a pattern of internal drainage to the northeast from 30 million years ago to 6 million years ago. At the end of that period, existing drainages were disrupted by the onset of tectonic extension, or stretching, and Basin and Range faulting to the west. This faulting produced low-elevation valleys that captured plateau streams and created the southwesterly flow pattern we see today in the Colorado River and its tributaries, including the San Juan River.

The geologic history of stream flow on the Colorado Plateau has been debated just as long as the history of uplift has. Early geologists looked at streams that cut across anticlines and surmised that the streams must have existed prior to uplift. During slow upward movement of the land surface, streams continued along their initial paths, essentially cutting through rising rock. As an analogy, imagine that instead of slicing down with a knife into a loaf of bread, you kept the knife stationary and lifted the bread. This theory of what happened here presupposes that features like the Monument upwarp formed recently and slowly enough that erosion could keep pace with uplift. If the uplift occurred more rapidly than erosion, it would have created a rock barrier, forcing streams in a different direction. But another, more likely hypothesis also explains how a river like the San Juan can go across, instead of around, an anticline.

When you look at the modern landscape at Goosenecks State Park, you're observing the result of millions of years of exhumation. There were most certainly younger Tertiary rock units that covered the older Paleozoic rock seen here. Some of those units were probably volcanic ashfalls and flows that filled large topographic irregularities left behind by the Laramide Orogeny. The internal drainage between 30 million and 6 million years ago filled other depressions with stream and lake deposits. All of this material is now gone, removed in the frenzy of erosion that followed Basin and Range faulting.

The looping meanders in the San Juan River are characteristic of streams that flow across broad, flat surfaces; although the modern topography here is obviously much different, this stream pattern is preserved, now entrenched deeply within solid rock. That entrenchment is the result of a rapid drop in base level, the lowest level to which any stream can erode. Most streams eventually spill into the sea. This is true of the three largest rivers in the world, the Amazon, the Nile, and the Mississippi, which empty into the Atlantic Ocean, the Mediterranean Sea, and the Gulf of Mexico respectively. Other streams, such as the Mojave River, empty into desert valleys and never reach the ocean. Traveling upstream from the mouths of rivers, streambed elevations increase. For rivers like the Nile and the Amazon, base level is sea level; for the Mojave River, on the other hand, base level is Death Valley. If sea level were to drop, streams that flow into the ocean would erode downward toward the new base level, and their tributaries would follow suit. In contrast, on the early Colorado Plateau, thousands of feet above sea level, lakes and depressions on the plateau itself controlled local base level since none of the streams here were connected to the ocean. When Basin and Range faulting suddenly opened this system to the ocean via the new Colorado River system, the main channel of the Colorado and its tributaries responded to this extreme base level reduction by rapidly eroding downward, creating incised channels like the one before you.

In the past 6 million years, the response of the plateau to erosion has been slow but continuous. As rock is removed from the surface, the Colorado Plateau thins and becomes lighter, resulting in an imbalance in the buoyant relationship between lithosphere and asthenosphere. To correct that imbalance, the lithosphere rises as the overlying mass decreases. This means that although sediment is continuously being removed from this region by stream erosion, the surface is not really lowering. Instead, it's thinning and rising to reestablish gravitational equilibrium. But as the surface rises, it elevates stream channels relative to base level (the Gulf of California), so streams incise more deeply in response to this disparity. Latest estimates by geologists place average uplift since late Mesozoic time, about 65 million years ago, at almost 7,000 feet, with 2,700 feet of rock removed by erosion in the last 30 million years. Exhumation of over 0.5 mile of rock has resulted in an isostatic uplift of 2,100 feet, almost one-third of the total uplift in the last 65 million years.

The Honaker Trail offers you an opportunity to visit the river. The trailhead is just west of the park and is accessible from the viewpoint road. The trail was built by miners who wanted access to marginal ore deposits below. Be aware that this steep, exposed path is not for the vertiginous. The first horse that was led down the trail fell to its death. And, in a classic case of "do as we say and not as we do," be careful about camping at the canyon rim. While writing this book, we spent a beautiful evening at Goosenecks State Park, eating a leisurely dinner and watching a brilliant sunset. Our pleasant evening left us unprepared for what followed, however, which was a torrential downpour accompanied by raging wind, blinding lightning, and crashing thunder. It was one of those nights where you lie spread out in an expanding puddle on the tent floor, trying to keep everything from flying off the canyon rim. There are probably better places to camp. . . .

Goosenecks State Park offers a view into the geologic history of uplift and incision on the Colorado Plateau. Here you see outcrops of rock that formed beneath the ocean exposed almost 1 mile above sea level due to continuing uplift of a large piece of the earth's surface. You also see a stream channel that has responded to uplift by cutting downward through rock to create a deep canyon in its search for the Colorado River and the sea. Remember that this is an ongoing process, as isostasy demands that for every grain of sediment removed, the surface must rise to reestablish equilibrium—imperceptibly in the short term, but immensely over geologic time. Be sure to visit Natural Bridges National Monument (see vignette 26) to the north to see how incised meanders like those at Goosenecks State Park create huge, impressive freestanding bridges of rock.

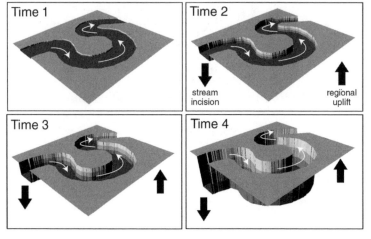

Uplift and incision on the Colorado Plateau

25 MESAS, BUTTES, AND SPIRES
Valley of the Gods

An ideal visit to Valley of the Gods begins with a drive through Monument Valley, and particularly Monument Valley Tribal Park in Arizona, just south of the Utah border. Seeing Monument Valley's majestic landscape and its famous outcrops such as Thunderbird Mesa, Wetherill Mesa, Mitchell Butte, East and West Mitten Buttes, Yei Bi Chei, and the Totem Pole, may put you in mind of classic Westerns like *Stagecoach*, which was shot here in 1938. Much of Monument Valley is part of the Navajo Indian Reservation, and access to secondary roads is restricted to visitors who have hired Navajo guides. Utah's Valley of the Gods, at the southern margin of Cedar Mesa, is north of Monument Valley, just over the San Juan River, and it features rock units and landforms similar to those found in Monument Valley, albeit on a smaller scale. Travel isn't restricted in Valley of the Gods as it is in Monument Valley, so visitors have more freedom to camp and hike. Both Monument Valley and Valley of the Gods are home to three categories of formations: mesas, buttes, and spires. Although geology textbooks give a rigorous definition for each of these words, the names attached to landforms in the real world aren't always consistent with those definitions.

To explore the geology of Valley of the Gods, start in the west and drive the 17-mile-long Road 242 through the valley from Utah 261 to U.S. 163. Although this graded road is suitable for passenger cars, pay attention to the weather, since occasional flash floods wreak havoc on roads like this. As you crest hills along this road, you'll be able to catch glimpses of the taller rock formations in Monument Valley in the distance to the south (right). Closer at hand, in the first 5 miles, you'll have nice views of Bell Butte to the south, and to the north you'll see Lady in the Bathtub with the prominent southern wall of Cedar Mesa in the background. Both formations feature horizontal bedding, but there are obvious differences. Bell Butte, at a lower elevation, has gently sloping sides with a very narrow top, while Lady in the Bathtub features vertical walls on a sloping base. Stop 1 is the top of a rise that provides you with views of Franklin, Rooster, and Setting Hen Buttes—this is a good place to talk about mesas, buttes, and spires. Pull over and park safely next to the road.

Mesas, buttes, and spires are all flat-topped, steep-sided features that form in arid landscapes with horizontal bedrock units. The flat tops are a function of horizontal bedding; if the bedding sloped, then so would the tops. The geologic definitions of these landforms are a function of the relationship between measured lengths. A mesa is wider than it is

Regional shaded relief map

GETTING THERE

From the Utah-Arizona border, drive 25 miles northeast on U.S. 163 through the small town of Mexican Hat to Utah 261 (see map, p. 190). Drive approximately 5 miles northwest on Utah 261 to the posted turnoff on the right for Valley of the Gods Road (Road 242). This graded dirt road proceeds east through Valley of the Gods from Utah 261 back to U.S. 163. Stop 1 lies 6.3 miles east of the Utah 261 turnoff; stop 2 is at mile 9.1; and you can pull over anywhere between mile 14 and mile 15 for stop 3.

tall, while a butte is taller than it is wide. A spire—sometimes called a tower, monolith, or monument—is much narrower than it is tall. If we look at these structures in the real world, we see that all three have sloping bases, vertical or near-vertical upper walls, and flat tops.

When exposed to weathering and erosion, stronger rocks maintain steeper slopes than do weaker rocks. In geology jargon, strong units are often called cliff formers, and weak units, slope formers. Shale is weaker than sandstone, so shale forms a sloping surface while sandstone faces often stand upright at a 90-degree angle. Mesas, buttes, and spires form where sandstone rests on shale, hence the common form of a vertical cliff standing atop a sloping base. Since shale erodes more rapidly than sandstone, the vertical sandstone walls are undercut. As this happens, pieces of sandstone break off from the sides and fall to form piles of blocky debris called talus. In Monument Valley, the steep-sided de Chelly

Cross section of a mesa, butte, and spire in Valley of the Gods

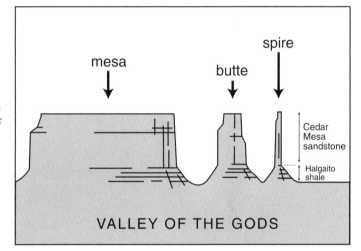

Looking east at Franklin Butte, Rooster Butte, and Setting Hen Butte

Strong Cedar Mesa sandstone, relatively resistant to erosion, rests on top of the thinly bedded, wavy Halgaito shale and creates unique rock sculptures like Lady in the Bathtub.

sandstone rests on a sloping base of Organ Rock shale, both Permian in age. Here in Valley of the Gods, the resistant Cedar Mesa sandstone lies atop the weaker Halgaito shale, also both Permian rock units.

Aridity is important in the formation of mesas, buttes, and spires because frequent rain, abundant vegetation, and rapid weathering act together to create gently sloping surfaces regardless of the underlying rock. Thus, vertical slopes are rare in humid climates, where weathering and erosion cause slope angles to become ever gentler, eventually creating a near-horizontal landscape. In arid regions underlain by horizontal beds, however, slope angles don't decrease at all; vertical slopes just retreat. Here in Valley of the Gods, removal of the underlying shale eventually causes the sandstone to fall, but the face of the cliff former remains nearly vertical. The landform shrinks in width while maintaining its height. Mesas are therefore precursors to buttes, which are in turn precursors to spires. When spires finally fall, they leave behind rounded mounds of eroding shale; Bell Butte, composed entirely of Halgaito shale, is just such a mound.

But what preceded the mesa that became a butte and then a spire? There's a somewhat hazy boundary between mesas and plateaus, which are expansive landscapes of high-elevation, flat-lying rock. The Colorado Plateau, which makes up large portions of the states of Utah, Arizona, Colorado, and New Mexico, is an obvious example of such a feature. But

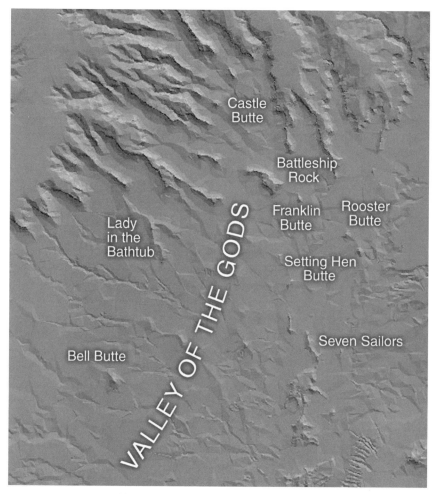

Shaded relief map of Valley of the Gods

the Colorado Plateau includes smaller plateaus separated from one another by faults, folds, canyons, and other defining features. Some surfaces that might be considered plateaus, such as Cedar Mesa, are called mesas. And yes, some features that are technically mesas are called buttes and vice versa. When is a butte not a butte? When the first person to name it calls it something else. History takes precedence over scientific rigor in geographic nomenclature.

The next 3 miles offer great views in all directions as you approach Castle Butte. Stop on the southwest side of Castle Butte; there's a convenient pullout at mile 8.5. This is a broad, imposing structure indeed. There's a surprise ahead, though. At mile 9.1 (stop 2), the road passes between de Gaulle and His Troops to the north and Castle Butte to the

From the west, Castle Butte appears to be a massive structure; from the north, its narrow fin shape, controlled by oriented fractures in the Cedar Mesa sandstone, is revealed.

south. What appeared from the west to be a massive butte is actually a very skinny ridge of rock called a fin, a term perhaps related to this type of structure's resemblance to a shark's fin slicing through ocean waves. We discuss fractures and fins in detail in vignette 29, on Arches National Park's Fiery Furnace. In general, fins are created when fractures isolate narrow ridges of rock. Here, a set of north-south trending fractures in the Cedar Mesa sandstone has produced this tall, narrow structure. All of the features in Valley of the Gods look different when viewed from different directions, and their colorful names typically make sense only from one angle.

At stop 3, you find yourself surrounded by talus, huge chunks of rock that have fallen from Rooster Butte and Setting Hen Butte. Some of these pieces are larger than automobiles. Looking up at the rock face, you can see light-colored scars that reveal where these boulders were before their precipitous descent. As we discuss in vignette 27, on Newspaper Rock, desert varnish covers exposed rock in arid lands. But when surface rock falls off, the true color of the bedrock is revealed. Rockfall is not a steady process. Instead, events like heavy rains or earthquakes cause sudden failure, accompanied by an explosively loud cascade of rock and dust. Frost wedging, a more continuous process, also leads to rockfall. Unlike most liquids, when water freezes, it expands. In winter, when water in rock fractures freezes, it widens the fractures. Daytime temperatures may allow this ice to melt, and the resulting water seeps deeper into the fractures to freeze and wedge once again. Over time, this repeated cycle eventually causes blocks of rock to fall away. Of course, erosion is a factor too, carving away the weaker underlying shale, subjecting the stronger Cedar Mesa sandstone above to the inexorable force of gravity. Eventually blocks of sandstone succumb to gravity and plummet to the ground; this is a form of mass wasting, the term for any gravity-driven downhill movement of rock or soil. As you continue to U.S. 163, examine Seven Sailors and Flag Butte. Note the difference between the slope maintained by the shale versus the sandstone layer. Can you see any evidence of recent rockfall? Are these features spires, buttes, or mesas?

Valley of the Gods offers an intimate glimpse into a landscape of mesas, buttes, and spires. These isolated structures, left behind by the eroding southern margin of Cedar Mesa, are constantly evolving. If you could return in 10,000 years, you'd see an altered landscape. Cedar Mesa will be slightly farther north, and Bell Butte, lacking a protective sandstone cap, will be considerably shorter. Perhaps gravity will have transformed Castle Butte or Battleship Rock into a tall, narrow spire, or maybe they will have lost their heads completely, becoming nothing more than rounded hillocks of shale. And undoubtedly there will be new and exciting buttes and spires emerging from the sandstone cliffs to the north to engage the imagination.

Talus at the base of Setting Hen Butte indicates ongoing mass wasting, the removal of rock by gravity.

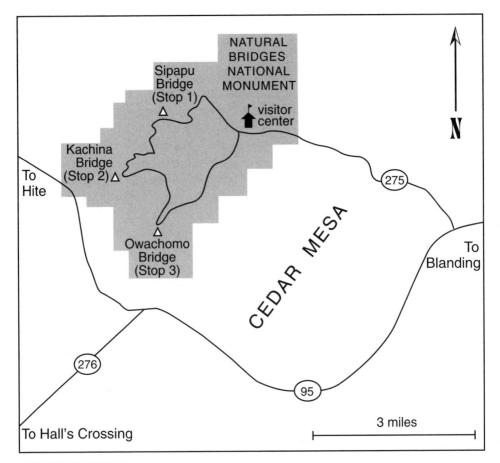

GETTING THERE

From Blanding, drive 4 miles south on U.S. 191 to Utah 95, then west 31 miles on Utah 95 to Utah 275. Take Utah 275 northwest 6 miles to the visitor center at Natural Bridges National Monument. From there, drive the one-way, counterclockwise loop (9 miles total) through the park, stopping and walking to each bridge in turn. The 1.2-mile round-trip trail to stop 1, Sipapu Bridge, loses 500 feet in elevation from the parking lot to the base of the bridge, and the Park Service classifies it as strenuous. You'll have to negotiate stairs and log ladders, so wear good shoes. The trail to stop 2, Kachina Bridge, is 1.5 miles round-trip and moderately strenuous, with steep sections of slickrock and a 400-foot elevation change. The trail to stop 3, Owachomo Bridge, is less taxing than the previous two trails, with no ladders or stairs. It's just 0.4 mile round-trip, with 180 feet in elevation change. If you prefer hiking to driving, an 8.7-mile loop trail also connects the three bridges. The Sipapu Bridge Trail is part of the loop trail. From Sipapu Bridge itself, the trail follows the sandy but flat canyon bottom to Kachina and Owachomo Bridges, then climbs up onto and across Cedar Mesa back to your starting point at the Sipapu parking lot. There is very little shade on the mesa top, so prepare for heat if you take the loop trail in summer. Whether you drive to the bridges or hike, take plenty of water.

26 BRIDGES ACROSS TIME
Natural Bridges
National Monument

There's some confusion about the difference between natural bridges and natural arches. Many people use the terms interchangeably; others are aware that the words describe different landforms, but they aren't sure of the exact definitions. Arches result from differential weathering and subsequent mass wasting, processes we describe in vignette 30, on Arches National Park. Natural bridges, like those here on Cedar Mesa,

Negotiating the trail to Sipapu Bridge

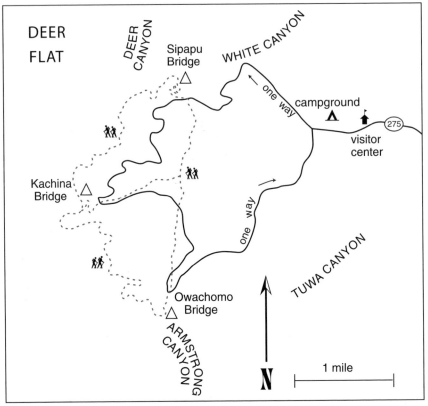

Natural Bridges loop road

form from stream erosion within deep canyons, and as their name implies, they occur as high spans from one canyon wall to the other. As we look at the different bridges in this vignette, keep in mind that this is an ongoing process. All of these bridges are continually reshaped by stream erosion and gravity. And, although similar, the processes that created Sipapu, Kachina, and Owachomo Bridges were not identical.

Let's begin our adventure at stop 1, the Sipapu Bridge trailhead. The names of all three bridges are derived from the Hopi language. Sipapu refers to the gateway to the world of spirits, a channel through which all souls enter and leave the world of the living. Sipapu Bridge, like virtually all of the visible rock in Natural Bridges National Monument, is made of Cedar Mesa sandstone from Permian time, formed from sand dunes with interbedded, or alternating, layers of stream and floodplain deposits. This is a sturdy, well-cemented unit, a necessary criterion for free-standing bridges.

As you work your way down to the bridge, you are in White Canyon, one of three drainages responsible for natural bridge formation on Cedar

Owachomo Bridge inspires Bob and Dave

Mesa. About halfway down to Sipapu, a rock ledge affords visitors an excellent view. You may choose to take photographs here, but be forewarned: the true size of the bridge is only apparent when you stand immediately beneath it. Sipapu is 220 feet high and spans 268 feet from canyon wall to canyon wall. The width of Sipapu is 31 feet at its narrowest point, about the same as a two-lane bridge. From the top of the bridge to the bottom of its unsupported span, known as its base, Sipapu is 53 feet thick at its narrowest point. Its size has led geologists to hypothesize that it is intermediate in age between the older, frailer Owachomo Bridge and the younger, sturdier Kachina Bridge.

In vignette 24, we talked about youthful streams, mature streams, regional uplift, and stream incision, all processes that contributed to formation of the Goosenecks of the San Juan River. Now we'll add another process to those already discussed. Young streams form in mountainous areas and flow in deep V-shaped valleys. Over time, stream erosion removes vast amounts of rock and creates muted landscapes. In these flatter landscapes, older streams flow in meanders, broad, sinuous loops that erode laterally, creating flat surfaces called floodplains. In a floodplain, the surface of the water in a channel is at virtually the same level as the surface of the land. This type of stream system can be re-energized by regional uplift like that experienced here and throughout the Colorado Plateau. Uplift creates high gradients in stream channels, providing energy for erosion. In a process called incision, streams then cut rapidly downward, and the broad loops that once sat on an almost flat surface become incised meanders, sometimes thousands of feet below a canyon's rim.

Meander cutoff, a process that takes place on floodplains, relates directly to natural bridge development. Every river has a thread of highest velocity flow, called the thalweg, which occupies the center of the channel along straight sections (*thalweg* is German for "valley way"). Perhaps you've waded across a stream and felt the tug of water increase toward the center of the channel and decrease toward the edge; what you experienced was the thalweg. When a stream flows around a meander, centrifugal force throws the thalweg to the outside in the same way your vehicle is thrown to the outside when you drive around a corner at high speed.

Sipapu Bridge is of intermediate age, appearing to be older than Kachina Bridge and younger than Owachomo Bridge.

In streams, high velocity equals high erosion and low velocity equals deposition. Wherever a stream flows around a meander, it erodes material on the outside of the loop (which is called a cutbank), and deposits material on the inside of the loop (called the point bar). Through this process, the channel moves outward. Sometimes two meanders move toward one another, and if they're close enough, the process of erosion gradually diminishes the space between them until water is able to breach it. Once that occurs, water takes the most efficient route, which is the steepest downhill path. The old channel is thereby cut off and bypassed, and sediment fills the gap between the active channel and the older channel. The cutoff loop becomes a stand-alone body of water called an oxbow lake, which gradually fills in with sediment.

How does this process relate to incised meanders like those on Cedar Mesa? Meander cutoff happens here, too. High flows in these desert canyons carry sand, gravel, and even boulders, all of which impact and abrade rock walls on the outsides of meanders. When these incised meanders erode outward and meet, though, they do so beneath high cliffs of resistant rock. The stream bypasses its previous course, flowing instead through a gap beneath rock that forms a natural bridge. Sipapu Bridge sits at a point in White Canyon where meanders have broken through the lower rock wall. The meander to the south was once part of the active channel, but now the stream flows due west beneath the bridge.

Stop 2 is Kachina Bridge, named after Hopi dancers and dolls with religious significance. This massive bridge is difficult to see from the

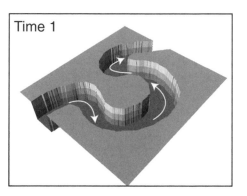

Time 1

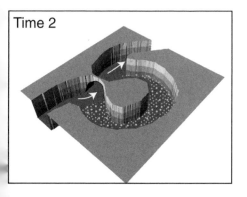

Time 2

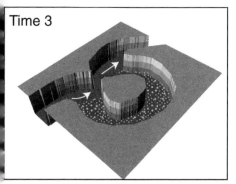

Time 3

Natural bridge formation and collapse in a canyon. The dotted area represents the previous path of the stream.

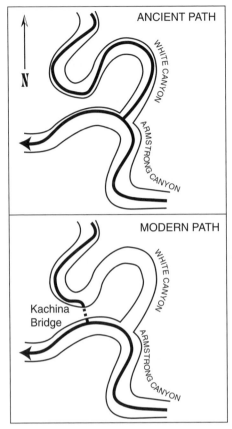

Formation of Kachina Bridge

trail. Standing beneath it, however, you can certainly see its broad shape— 210 feet high with a 204-foot span. Kachina is 44 feet wide on top, and at 93 feet thick, it's almost twice as thick from top to base as Sipapu. Kachina formed from the interaction of two converging drainages, White Canyon from the north and Armstrong Canyon from the east. White Canyon originally followed a southerly meander, then a northerly meander prior to joining Armstrong Canyon. The southerly meander, however, eroded outward, eventually breaking through the base of the rock wall to create hulking Kachina Bridge above. The northerly meander became a dry and abandoned channel.

Geologists believe Kachina is the youngest of the three bridges. Like most landforms, natural bridges evolve over time. Once a bridge forms, gravity is its enemy. All rock is fractured to some degree, and new fractures form as a function of weathering. Rock weakened by fractures eventually succumbs to gravity and falls. When this occurs on a natural bridge, its base diminishes, growing thinner. In the summer of 1992, 4,000 tons of rock fell from the base of Kachina Bridge, thinning it and widening the gap beneath it; you can still see an immense pile of boulders lying on the northern side of the channel. Over time, a bridge grows increasingly delicate. Eventually it fails completely in a final, grand collapse. We can't assign absolute ages to the bridges on Cedar Mesa, but we can guess their relative ages based on how robust or delicate they appear.

Stop 3 is Owachomo, which means "rock mound" in Hopi. The oldest and frailest of the three bridges, it's 27 feet wide on top, but only 9 feet thick at its narrowest point from top to base. It stands 106 feet high and stretches 180 feet across the canyon. As at Kachina Bridge, two drainages interacted to create this bridge, but the story here is bit more complicated. Tuwa Canyon and Armstrong Canyon both head west, with Tuwa Canyon lying north of Armstrong Canyon. The original junction of the canyons, west of Owachomo Bridge, has been abandoned due to a southerly meander breaking through the south wall of Tuwa Canyon and breaching Armstrong Canyon. It was this cutoff that created Owachomo Bridge long ago; estimates of its age vary from 100,000 to 30,000 years old. But this channel is abandoned also, as another southerly meander to the east of this one also broke through into Armstrong Canyon, perhaps creating a new bridge in the process. If there once was a bridge over the active channel, it doesn't exist anymore—perhaps it has fallen to earth and been swept away.

Geologists see around them a dynamic world that never stops evolving. Here at Natural Bridges National Monument, we have the opportunity to observe snapshots in time. Kachina is a young bridge, the result of recent stream erosion and undercutting. Sipapu is mature, standing tall and strong over an active channel. Owachomo Bridge is the frail, elder landform of Cedar Mesa; it has long stood as a lonely sentinel surveying a waterless channel. When you look at these magnificent rock spans, remember that as they fracture and fall, new bridges are being created elsewhere in the park. As long as water flows in these deep canyons, erosion will continue to cut away at the outer margins of meanders, eventually breaking through to create new bridges.

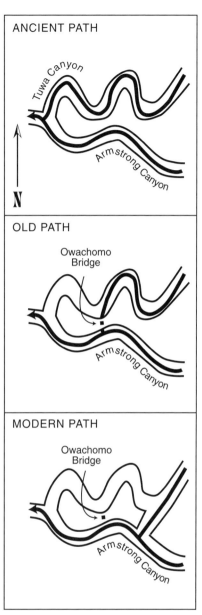

Formation of Owachomo Bridge

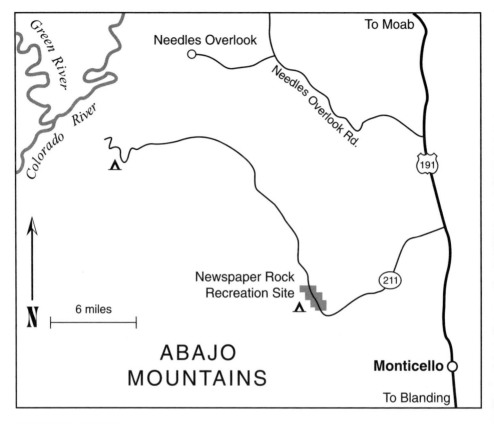

GETTING THERE

Newspaper Rock Recreation Site, overseen by the Bureau of Land Management, is east of the Needles District of Canyonlands National Park in the foothills of the Abajo Mountains. From Moab, take U.S. 191 south 40 miles to Utah 211. A prominent geologic formation, Church Rock, is at the junction on the east side of U.S. 191. From Monticello, take U.S. 191 north 14 miles to that same junction. Newspaper Rock is 13 miles west on Utah 211; the turnoff is clearly signed. The rock itself is about 50 feet from the parking lot. Do not cross the protective fencing around the rock.

27 TSÉ HANE, THE ROCK THAT TELLS A STORY
Newspaper Rock

Newspaper Rock, a rock art panel about 15 feet tall and 20 feet wide, sits in a shaded alcove at the bottom of a tall sandstone face near Indian Creek. Incised images—hundreds of them—cover the black, steeply sloping panel, and a natural roof above it protects the art from weathering and erosion. The pictures on Newspaper Rock are petroglyphs, images carved into stone. In contrast, pictographs are painted onto rock surfaces, not inscribed; there are no pictographs at Newspaper Rock.

Newspaper Rock sits in a sheltered alcove that helps protect the petroglyphs from weathering.

The Navajo name for Newspaper Rock is *Tsé Hane*, "the rock that tells a story." Actually, it seems to tell many stories; the sheer number of images on the panel can be overwhelming. Take your time and look carefully at individual images, which include anthropomorphs, or human-shaped figures. In the upper center of the panel, a particularly striking anthropomorph with horns and carrying a knife appears among a group of images that have taken on a heavy patina. Four-, five-, and six-toed footprints and paw prints are common, as are series of tracks. The panel abounds with images of mountain sheep, deer, and other quadrupeds, and some of the prints and tracks are cloven. There are also at least four sets of concentric circles, as well as other abstract circular and curving symbols. As you study the panel, notice the darker images below those on the surface; these are even older petroglyphs. Together, these darker images and the lighter ones atop them represent over two thousand years of artistic expression.

Who carved these figures, and what do they mean? We'll get to that, but first let's start where the artists did, with the canvas itself—in this case, desert varnish. Notice the profound difference in color between the deep black panel and the much lighter weathered sandstone above and beside it. Petroglyphs emerge as an artist with a rock tool pecks through the dark surface layer, exposing the lighter-colored rock underneath.

Rock varnish clings to an unweathered portion of this sandstone face.

The thin, dark surface layer, called desert varnish or rock varnish, is ubiquitous on rock in arid and semiarid regions. Researchers originally believed this black to reddish brown coating precipitated or leached from underlying rock by means of weathering. However, further analysis has shown that the high levels of manganese and iron found in desert varnish are absent in many underlying rock types.

It turns out that the mysterious origins of desert varnish are in part biological. Rock surfaces in arid regions host colonies of microbes. These bacteria, for example, *Arthrobacter* and *Metallogenium*, absorb manganese and iron from the air and use them in metabolic functions, just as we use food. The microbes secrete manganese oxide and iron oxide as waste products, forming a sticky film on rock surfaces. The more manganese oxide secreted, the blacker the coating; the more iron oxide, the more orange the coating. Clay, which composes up to 70 percent of desert varnish, arrives as windblown dust that adheres to the sticky manganese-iron coating. From then on, the clay acts as a protective barrier against the

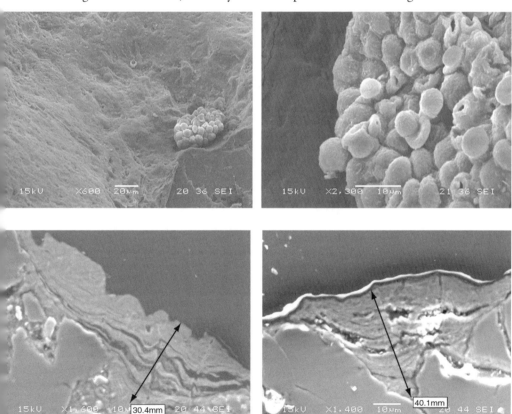

Top L and R: *Rock surfaces in arid regions host colonies of microbes that aid in desert varnish formation.* Bottom L and R: *Varnish forms as a thin patina of clay materials, manganese oxides, and iron oxides.* —T. Diaz, scanning electron microscope photo

The surface of Newspaper Rock is covered with petroglyphs
of a variety of ages and cultural affiliations.

elements. Desert varnish layers are exceptionally thin, typically only 10 to 100 microns thick—that's 0.1 to 0.01 millimeter, or less than one-fifth the thickness of a standard sheet of paper. It accumulates very slowly, averaging about 10 microns per thousand years, and exposed varnish is constantly assaulted by the elements. If you look at the sandstone walls around Newspaper Rock, you'll see small patches of varnish on largely unvarnished faces. Where varnish is lacking, weathering has altered and weakened the rock until it crumbled and broke away, taking with it the surface patina.

Researchers are still refining techniques for dating desert varnish precisely, but for now, the relative darkness of the patina can be used to assign relative ages to petroglyphs. For example, older petroglyphs, revarnished over time, are darker and less distinct against the varnished background, so the older the carving, the darker its color. Younger petroglyphs, on the other hand, are lighter and more visible. The layering of images confirms this hierarchy. The lightest petroglyphs always cut across darker petroglyphs; it's never the other way around.

Humans have inhabited Utah for nearly 12,000 years. The 7,564 known rock art sites in the state were created primarily in the last 8,000 years. Newspaper Rock is unique among these sites because artists from at least four Native American cultures carved the images here. In most

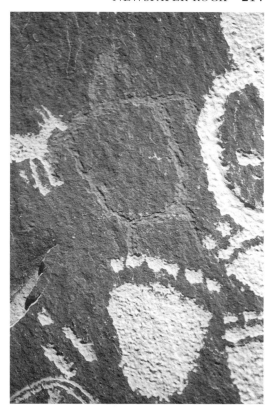

An older anthropomorph is mostly revarnished, while younger petroglyphs appear very fresh.

cases, we can't tell exactly which culture created which image on the panel. But with the information we provide about each culture's artistic style, see if you can make some educated guesses.

The earliest known artists represented at Newspaper Rock belonged to the Western Archaic culture. These were nomadic hunters and gatherers who prospered from 9000 BC to about 1 AD. In general, their style features shamanistic or supernatural anthropomorphs, circular symbols, and abstract, curving lines. Western Archaic style influenced the rock art of later cultures as well. Some of the most impressive Western Archaic rock art in this area is near the Maze District of Canyonlands National Park. In two sites in Horseshoe Canyon, broad-shouldered anthropomorphs, some of them over 6 feet tall, parade in the Great Gallery. The earliest confirmed petroglyph in Utah—8,700 years old—was found on a stone in Cowboy Cave, up the canyon from the Great Gallery.

The Ancestral Pueblo culture thrived across southern Utah and in the Four Corners area from 200 BC until the arrival of the Spanish in 1540 AD. (The term Ancestral Pueblo is now preferred over the more familiar name Anasazi for a variety of reasons.) Ancestral Pueblo rock artists seem to have been active from 1 AD to 1300 AD, and what is now the Needles District of Canyonlands National Park became a prominent northern Ancestral Pueblo settlement around 850 AD. Ancestral Pueblo rock art styles in this region include anthropomorphs with triangular, trapezoidal, tapered, or sticklike bodies, mainly standing with arms out to the sides and fingers and toes spread, if shown. An Ancestral Pueblo artist probably carved the horned anthropomorph with the knife mentioned earlier. Researchers theorize that horns, whether on anthropomorphs or animals, represent godlike or supernatural powers. Later Ancestral Pueblo rock art styles include humpbacked flute players; anthropomorphs with facial features and elaborate headdresses; bighorn sheep and other quadrupeds, often with open mouths; and spirals, concentric circles, and prints of hands, feet, and paws. Experts believe paw prints were symbols of power in some prehistoric cultures. In one interpretation of concentric circles in Ancestral Pueblo rock art, the outermost circle is the light around the sun, the second circle is the sun itself, and the innermost circle is the sun's umbilical connection, which provides sustenance to all living things.

Fremont culture dates from 450 to 1300 AD. The Fremonts' realm covered the majority of Utah north and west of the Colorado River and small areas in surrounding states. Within that area, Fremont rock art style varied; the Southern San Rafael style is the one most likely to be found at Newspaper Rock. In this style, anthropomorphs have horned or ornamented headgear, sharply angled and tapered torsos, feet pointing to the sides, and facial features, including slit eyes. Mountain sheep were important to the Fremont and seem to have both supernatural and utilitarian meaning in their art. Fremont mountain sheep often have crescent-shaped or square bodies and thin horns. Images of shields, snakes (also symbols of power in some cultures), and animal tracks, as well as abstract line patterns, also define the Fremont style.

Hunting scenes on Newspaper Rock depict riders with bows on horseback. These are undoubtedly the work of the Ute, who replaced the Fremont in this area after 1250 AD. Like earlier styles, Ute rock art features anthropomorphs, quadrupeds, and handprints and paw prints, but it's also distinguished by such subjects as guns, bison, teepees, and especially horses. Rock art with horses (and guns) in it must have been produced after 1540, when the Spanish reintroduced horses on the continent after thousands of years of extinction. Most lighter-shaded petroglyphs at Newspaper Rock have been attributed to Ute artists between 1830 and 1880. Look for bison on the panel and a figure with a quirt and fringed leggings or chaps, the latter suggesting a Euro-American influence on the artist's culture. In some places, you can also see where Utes drew over previous work, either repecking old petroglyphs or covering them with new art. This complicates attempts to assign relative age or cultural affiliation to individual images.

Why did these artists carve images on Newspaper Rock? While Hopi, Zuni, and other modern-day descendants of these ancient peoples have provided some clues, there is no Rosetta stone for petroglyphs; thus, understanding their meaning and cultural origins is difficult. But theories abound: Some say the artists were hunters and that the petroglyphs were part of ritual ceremonies for bringing animals within their reach. Others associate rock art with vision quests or shamanistic activities, such as initiations. Styles and figures were often passed on or borrowed from culture to culture, and within cultures, styles evolved with time.

Petroglyphs are fragile cultural resources vulnerable to destruction and vandalism. You can see before you what vandals have done to Newspaper Rock, and similar scrawls can be seen at other sites across southern Utah. One truly has to wonder what the people who perpetrated these acts got out of it. Both the Antiquities Act of 1906 and the Archaeological Resources Protection Act of 1979 protect rock art from malicious intent, and vandals are subject to fines and imprisonment. Don't even touch the rock face innocently or out of curiosity. Oils on human skin can discolor rock and damage rock art. Treat these panels with the same respect you'd show any great work of art in a museum.

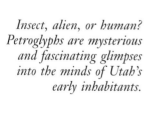

Insect, alien, or human? Petroglyphs are mysterious and fascinating glimpses into the minds of Utah's early inhabitants.

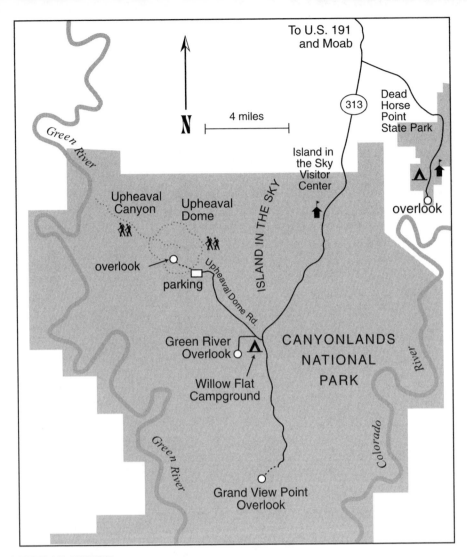

GETTING THERE

Upheaval Dome is located in the Island in the Sky District of Canyonlands National Park. To reach the Upheaval Dome Overlook from I-70, take exit 180 and travel 21 miles south on U.S. 191 to Utah 313. Or, from Moab, take U.S. 191 north 9 miles. Turn west on Utah 313 and go 22 miles to the Island in the Sky Visitor Center. On the way, you'll pass the turnoff to Dead Horse Point State Park. From the visitor center, it's 6 more miles on Utah 313 to the junction with Upheaval Dome Road (Road 183). Turn right on this road and go 5 miles to the overlook parking lot. There are several options for exploring this site. From the overlook parking lot, you can take a short trail to view the core of Upheaval Dome. The very challenging 8-mile round-trip Syncline Loop Trail takes off from a junction with this short trail. Upheaval Canyon Trail, off White Rim Road, is 1 mile shorter with a more gradual elevation change than the Syncline Loop Trail, but the trailhead is not at this location. Details about these trails follow in the text. Summer temperatures on these trails often exceed 100 degrees Fahrenheit, so if you hike, be sure to take enough water (we actually ran out) and, if possible, start early in the morning.

28 AT THE MYSTERY'S CORE
Upheaval Dome

When you pull your car into the Upheaval Dome parking lot, you're already within the outer ring of a mysterious geologic structure. Take the short trail to the Upheaval Dome Overlook. Before you is a multi-ringed structure 3.4 miles in diameter. You are standing on an inner ring of resistant Wingate sandstone from Triassic time. Behind you, a valley is carved into the less resistant early Jurassic Kayenta formation, and beyond that is a resistant outer rim of Jurassic Navajo sandstone (this last rock type is the subject of vignette 8). In front of you is a basin, a massive crater made of the weaker Chinle and Moenkopi formations. In the basin's core, slabs of White Rim sandstone stand upright. Barely visible between the white sandstone beds is the underlying Organ Rock shale. Here, *dome* refers to the upward curvature of layers, something you'll note by looking around the inner rim of this circular structure. Upheaval Dome's striking appearance attests to a dramatic episode in the distant past. For a better sense of the scale of the entire ring structure, you may wish to hike back toward the parking lot and take the trail up to the Navajo sandstone ridge presently behind you, but the best view of the core is right where you are.

There's no question about the beauty and uniqueness of Upheaval Dome. However, since the 1920s, geologists have argued vehemently about most other aspects of the site—in particular, its origin. Is it a remnant of a strange volcanic eruption? A salt intrusion? The aftermath of an ancient meteorite impact? Studies support each of these theories, but more recent evidence points to a single cause.

221

Panoramic shot (top left and right) *of Upheaval Dome*

Hiking the inner ring, upon which you stand at the overlook, will reward you with different views of Upheaval Dome. But if you're an experienced hiker, you can get a closer look on the Syncline Loop Trail, which follows the perimeter of the dome structure. This 8-mile loop trail includes two very challenging and steep descents and ascents across uneven rockfall debris and totals 1,400 feet in elevation change. About halfway around the loop, another trail takes you 1.5 miles (one way) into the dome's core. You can also access this core trail by the White Rim Road, a 111-mile dirt road that loops around the Island in the Sky area. To get to that trailhead from the overlook, drive back to Utah 313 and go north. Look for Mineral Road, also known as Horsethief Trail, on the left 8.2 miles past the Island in the Sky Visitor Center. From the junction of Utah 313 and U.S. 191, Mineral Road is on the right approximately 14 miles down Utah 313. On Mineral Road, drive 12.9 miles to a junction, turn left, then proceed another 7 miles to the Upheaval Canyon trailhead. This trail is 3.5 miles (one way) to the core trail. There's a small campsite near the junction of the Upheaval Canyon Trail, the Syncline Loop Trail, and the core trail.

One of the early theories of how Upheaval Dome formed involves volcanic activity. This cryptovolcanic theory suggests that gases from a

Upward-tilting sandstone beds surround the central core of Upheaval Dome.

Trail to the core of Upheval Dome, from Syncline Loop Trail

magma body exploded out of the earth's surface, forming the crater; it also maintains that no igneous rock intruded the structure or erupted from it afterward. (*Crypto* means "hidden" or "secret"; *cryptovolcanic* refers to the presence of features associated with volcanic activity at sites, such as Upheaval Dome, that lack any clear evidence substantiating volcanic activity.) There are many examples of igneous activity in this region, for example, in the La Sal Mountains to the east (see vignette 32). For a time, cryptovolcanism was the main hypothesis geologists used to explain large craters associated with catastrophic scenes such as the one you see here. Iowa's Manson Impact Structure is a similar site. However, the cryptovolcanic theory has been discarded at Upheaval Dome, and at other sites, based on subsequent research. That leaves salt intrusion and meteorite impact as the processes in competition for having created Upheaval Dome.

Upheaval Dome is located within the broad expanse of the Paradox Basin, an ancient ocean basin that formed during Pennsylvanian time alongside the southern border of the Ancestral Rocky Mountains, also known as the Uncompahgre uplift. As the uplift rose, ultimately to estimated elevations of 12,000 to 15,000 feet above sea level, this basin, adjacent to it, sank. A rim near sea level on the basin's southwestern

Three-dimensional image of Upheaval Dome

boundary acted as a spillway and, combined with tectonic and climatic changes, caused alternating periods of flooding and evaporation of saline seawater. This produced the Paradox formation, consisting mostly of salt with small quantities of other marine sediment, which accumulated to an exceptional thickness—over 5,000 feet at its deepest near the Uncompahgre uplift. Salt has a very low density, and as other sediments accumulate on top of it, the weight of the overlying material above causes the salt to deform and rise in mushroomlike mounds called diapirs. The buoyant salt diapirs force their way through overlying layers, leaving them tilted upward toward the path taken by the salt. Is Upheaval Dome part of the vertical path taken by a salt diapir as it rose to the surface? Let's look at the competing hypothesis before trying to answer that question.

First of all, what distinguishes a meteor from a meteorite? A meteor is a body of rock flying through outer space that enters earth's atmosphere at incredible velocity (usually measured in thousands of miles per hour). Most meteors burn up when they enter our atmosphere due to frictional heating. "Shooting stars" aren't really stars; they are meteors. A meteorite is a meteor that survives its encounter with the atmosphere and strikes the earth's surface. The meteorite impact hypothesis for Upheaval Dome suggests that a body of rock struck the surface above the dome (remember

Erosion has left a deep valley that surrounds the central core.

that rock layers that used to overlie the dome have been eroded), creating the subsurface structure you see today. The dome is actually the bottom of the impact point. Some geologists believe this happened in Jurassic time. Nearby outcrops of the Jurassic Carmel formation, long eroded away from this site, have unique soft-sediment deformation features that indicate regional disruption on a grand scale. These features, evidence of significant shaking before hardening, include wavy beds, evidence of shearing, and pipes, sand-filled tubes created through rapid liquefaction. If these features formed in soft sediment, that means the Carmel formation was not yet cemented into solid rock. If an impact occurred during this time, geologists estimate that 300 to 600 feet of the dome would have since eroded away. However, the timing of an impact is still being debated.

There are three stages to a meteorite impact: contact and compression, excavation, and modification. The meteorite's initial contact with the earth lasts between 0.1 and 0.001 second, depending on its size and speed. In the second stage of impact, a shock wave moves outward from the contact point and excavates a large crater. The third stage entails collapse of unstable edges of the crater and core rebound. To understand the latter, consider what happens when you drop a stone into a pond.

A large depression forms, then collapses again. When it has collapsed completely, the center rebounds and water spouts up in the center exactly where the stone disappeared underwater. There are several potholes at the Upheaval Dome Overlook. If it has rained recently, take a moment to observe this dynamic in action. Drop a pebble in the water and watch what happens.

What evidence exists for the impact theory? Erosion has removed any remnants of the meteorite, but there are shattered sand grains in the dome's core as well as shatter cones—conical fractures produced by shock waves—in the Moenkopi formation there. Both of these features occur at known impact sites, and shatter cones have also been found after nuclear explosions. The dome itself is not uniform; it has upwarps and downwarps, one of which gives the Syncline Loop Trail its name. Additionally, sets of thrust faults indicate horizontal stresses possibly relating to the excavation and modification phases of impact.

Rock in the very center of Upheaval Dome stands nearly vertical.

Could salt have created Upheaval Dome? Although salt is present below Upheaval Dome, there's no evidence of salt or other marine sediment in the core. There are diapirs in the Paradox Basin, but none of them look anything like Upheaval Dome. Not only that, according to some geologists, no diapir in the world has a structure similar to Upheaval Dome. Lastly, movement of a salt diapir probably could not have produced the shattered sand grains, shatter cones, or horizontal shear structures found in Upheaval Dome.

What's the answer? Though some disagreement continues, the most recent studies point to the meteorite theory. Although friction between atmospheric gases and space debris can raise a meteor's temperature enough that it burns up before impact, the rock that created Upheaval Dome was too big to burn completely. It struck the earth's surface with the force of an atomic bomb. Any dinosaurs that saw its arrival firsthand were vaporized by the energy released on impact. A huge crater formed, then part of it collapsed as the central core rebounded upward into the sky. Rock and dust probably continued to fall for days or weeks, then all was finally still. Eventually plants and animals recolonized the area and millions and millions of years passed, until at last you showed up here, peering quizzically into the core of the mystery that is Upheaval Dome.

29 A SEA OF FINS
The Fiery Furnace

The Fiery Furnace in Arches National Park is a convoluted maze of aligned rock ridges, called fins, separated by narrow, deeply eroded divides. Devils Garden, to the northwest in the park, is another such area, but the Fiery Furnace has passageways that are more tightly constricted, drop-offs that are steeper, and more hidden pools. It is such a labyrinth, in fact, that the Park Service assigns rangers to guide visitors along its confusing paths, and for good reason: tourists have become hopelessly lost in the Fiery Furnace, often within 1 mile of the parking lot. So if you have any doubt about your navigation skills, visit this area with a group. If you prefer to see it on your own, the park requires that you sit through an orientation video that stresses safety.

Narrow canyons, steep slopes, and hot summer temperatures make travel within the Fiery Furnace somewhat risky. Visitors have gotten lost, often very close to the parking area.

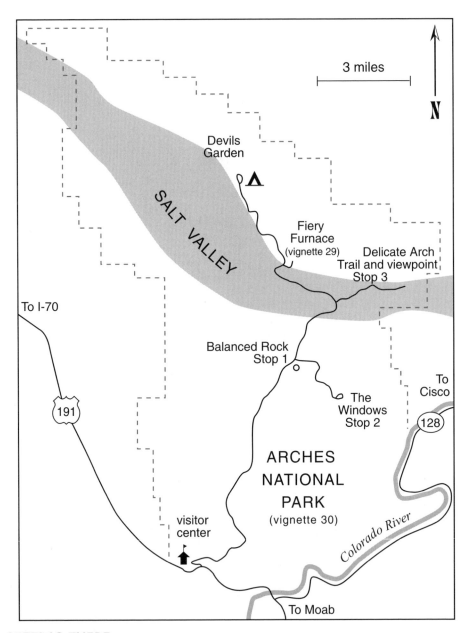

3 miles

N

Devils
Garden

Fiery
Furnace
(vignette 29)

Delicate Arch
Trail and viewpoint
Stop 3

SALT VALLEY

To I-70

Balanced Rock
Stop 1

To
Cisco

191

The
Windows
Stop 2

128

ARCHES
NATIONAL
PARK
(vignette 30)

Colorado River

visitor
center

To Moab

GETTING THERE

From exit 180 on I-70, drive 26 miles south on U.S. 191 to the entrance of Arches National Park. From Moab, drive 5 miles north on U.S. 191 to the park entrance. From the visitor center at the entrance, go approximately 14 miles along the main park road to the marked turnoff to the Fiery Furnace parking area. Several trails into the Fiery Furnace start here, but read the safety precautions that follow.

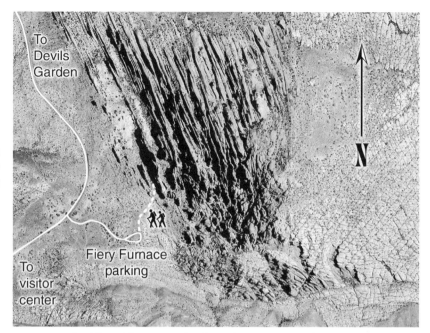

Aerial photo of the Fiery Furnace

As you enter the Fiery Furnace, the first thing you're likely to notice is how tightly constricted the paths are that wind back and forth beneath the towering sandstone fins and spires. Take some time to appreciate the forms you see here. All are carved into the Jurassic Entrada sandstone, which is also responsible for the many arches in the park. In fact, fins and arches are related structures, fins being the predecessors that lead to later arch development. We discuss the process whereby fins become arches in vignette 30. You've probably noticed by now that the fins here appear to be aligned with one another, all oriented from southeast to northwest. Interestingly, that's also the orientation of Salt Valley, the broad depression to the south, and the Uncompahgre Plateau to the north. This parallel orientation is no coincidence.

If you're like us, one hand is holding this book while the other hand periodically dips into a bag of tasty chips or pretzels. One reason they taste so good is their high salt content; unsalted potato chips just aren't the same. All of us consume salt, and our modern diet contains much more salt than did historic and prehistoric diets. But where does all that salt come from? The ocean, you say? Small quantities of salt, specifically sea salt, are produced by evaporation of ocean water, but the vast majority of salt comes from large mines that tap into thick underground deposits. One such mine is currently removing salt beneath Lake Erie, one of our Great Lakes; workers in Cleveland, Ohio, travel 2,000 feet below the earth's surface daily to remove 400-million-year-old salt deposits. Salt is, after all,

This delicate tower stands as the eroded remnant of a fin.

a mineral like any other, and minerals are the building blocks of rocks. But if you said, in response to our question above, that salt comes from the ocean, you were correct after all, because salt that lies underground in the modern world was once dissolved in ancient oceans.

In late Paleozoic time, 300 million years ago, the region that is now Utah sat very near the equator. A basin encompassing present-day southeastern Utah and southwestern Colorado sank just west of a northwest-trending fault to create an inland sea. To the west of the fault, a young mountain range was just beginning to rise—the Ancestral Rocky Mountains, also called the Uncompahgre uplift. The Paradox Sea, as this ancient body of water has come to be known, experienced a series of unique conditions that led eventually to the accumulation of huge quantities of precipitated salt. The Paradox Sea was alternately connected to the world's oceans, then cut off from them by changes in water level and topography. The prevailing warm, arid climate caused complete evaporation of the inland waters during times of isolation. Evaporation causes precipitation of minerals in solution, and for the Paradox Sea, the most abundant dissolved mineral was halite, known to most of us as table salt.

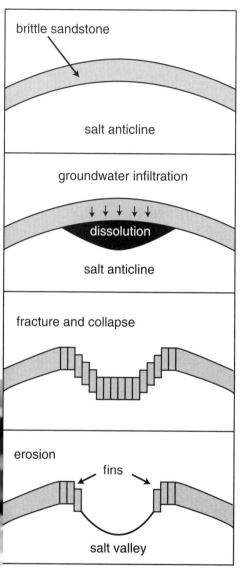

*Subsurface dissolution and collapse
lead to fin formation.*

Hence, a single cycle of connection, isolation, and evaporation resulted in a thick bed of salt on the land surface, much like that seen in the Great Salt Lake Desert, a remnant of the former Lake Bonneville (discussed in vignette 5). But there were many cycles of filling and evaporation in this sea—twenty-nine in all, over a period of about 15 million years. As more and more salt precipitated, increased mass made the basin sink ever deeper. When the Paradox Sea left for good, its signature remained in 6,000 vertical feet of salt—the Paradox formation—spread out across 12,000 square miles.

Ongoing growth of the Ancestral Rocky Mountains peaked about 275 million years ago. Stresses associated with uplift of that range caused Paradox formation salt to flow. Though it may seem surprising that salt can flow, it is a plastic material, one that exhibits characteristics of both fluids and solids. Silly Putty, the child's toy, is a great example of such a material (geologists play with Silly Putty too). If you place Silly Putty on a sloped surface, it will flow downward; hence, it is obviously a fluid. But if you pull hard on it, it will break. So it's really a solid. In reality, it's something in between: a plastic material. Here in southeast Utah, the most active period of mountain growth correlated with the period of highest regional stress; therefore it also correlated with the most salt movement. In some places, the salt thickened and pushed overlying rock up from below, creating anticlines, or upward folds in rock (for more information about anticlines, see vignette 6). In other places, the salt thinned. Some of the anticlines once featured salt cores over 10,000 feet thick! Later flow was much less dramatic. All of these anticlines developed a consistent

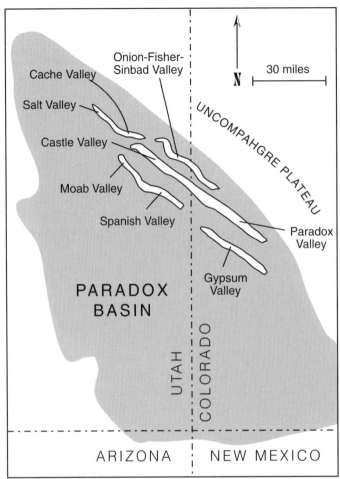

Regional collapse valleys along salt anticlines

Preferential erosion along joints has created deep canyons between sandstone fins.

northwest-to-southeast orientation, perpendicular to the applied stress. We now jump to the last 10 million years of earth history and the rise of the Colorado Plateau. As discussed in vignette 24, uplift of the plateau was accompanied by downcutting and removal of a huge volume of rock. As material was removed, surface water infiltrated into older and older rock units. Eventually water entered the Paradox formation about 2 million years ago. We all know what happens when you dump salt into a glass of water: it dissolves. When water moves into a formation like the Paradox, the salt begins to dissolve, leaving cavities below the surface. Dissolving anticlines left huge voids, which then collapsed into parallel valleys, including Salt Valley (seen here in Arches National Park), Cache Valley, Castle Valley, Moab Valley, and Spanish Valley.

Brittle overlying rock units like the Entrada sandstone fractured as valleys fell inward. The fractures, called joints, formed parallel to valleys and from the center outward as the valleys widened. What you see in the Fiery Furnace are those joints; they make up the spaces you meander through between tall fins, which are remnants of unfractured rock between joints. Joints allow water to percolate more easily into rock, which increases the rate of weathering and widens avenues between standing rock. One of the key differences between Devils Garden to the north and the Fiery Furnace is how much joint widening has taken place. Joints at the Fiery Furnace are still relatively tightly spaced, making them difficult to navigate, while joints at Devils Garden are wider, creating more accommodating passageways. The joints are probably wider in Devils Garden because these joints are older and have experienced more weathering and erosion. Fiery Furnace is still composed more of rock, whereas Devils Garden is more joint. All of these joints are aligned in the same orientation as the original salt anticlines and collapsed valleys: northwest to southeast. Remember this pattern; it's a useful aid to navigation as you explore the Fiery Furnace.

One of the most interesting things about any landscape is the long and seemingly improbable sequence of events responsible for surface landforms. Here at the Fiery Furnace, you see the result of a 300-million-year-old process that begins with an ocean and ends with uplift of the Colorado Plateau, ultimately forming this unique landscape of tall fins and tight passages. That uplift has done more than just create sandstone landforms, however; it has also liberated long-trapped minerals. When water flows through Salt Valley, it picks up salt originally deposited by the drying Paradox Sea. Streams carry this salt to the Colorado River, which flows southwest then due south into the Sea of Cortez. After hundreds of millions of years, salt from an ancient ocean is returned once more to the sea.

30 WEATHERING THE TESTS OF TIME
Arches National Park and the Entrada Sandstone

Arches National Park is home to thousands of rock arches that vary in size from tiny peepholes to the huge Landscape Arch, which is over 300 feet wide and 100 feet tall. While sandstone landforms of all shapes and sizes decorate the landscape here, arches are the hallmark of this park. To form arches, rock must be both strong and weak; it must weather but it also must be able to stand in fins, tall structures shaped like the dorsal fins of fish. Arches rarely occur in humid regions, as rock there disintegrates and decomposes too rapidly. In Arches National Park, which lies in the arid Colorado Plateau, the perfect combination of rock type, rock structure, and climate has created a magnificent assemblage of arches.

GETTING THERE
From exit 180 on I-70, drive 26 miles south on U.S. 191 to the entrance of Arches National Park (see map on p. 230). From Moab, drive 5 miles north on U.S. 191 to the park entrance. Stop 1, Balanced Rock, is 9.2 miles north of the visitor center on the main park road; there's a parking area on the right. Stop 2 is the Windows trailhead. To get there, turn right (east) off the park road just north of Balanced Rock onto the paved road leading to Double Arch and the Windows Arches. Park and take the 1-mile loop trail to North Window, South Window, and Turret Arch. For stop 3, Delicate Arch, go back to the main park road; turn right (north) and go 2.5 miles to the turnoff to Wolfe Ranch. Turn right (east) and drive 1.2 miles to the parking lot for the viewpoint and trailhead. The Delicate Arch Trail is 3 miles round-trip. Other than a few sections with exposure to heights, the trail is only moderately difficult, with a 480-foot elevation gain.

We'll begin our discussion of arch formation at stop 1, Balanced Rock, an immense sandstone boulder resting on a tall, narrow base. If you want to get a closer look, take the short, 0.3-mile loop trail around the base of the rock. Balanced Rock is the result of differential weathering. The term weathering covers all processes involved in the physical and chemical breakdown of rock exposed at the earth's surface. While many weathering processes act simultaneously to break down rock, we'll focus on two that are more important here: frost wedging and dissolution.

Frost wedging results from a rather unique characteristic of water: unlike most substances, water takes up more space in its solid form (ice) than it does in liquid form. Consider a typical winter day in Arches National Park. By early afternoon, the sun has warmed the surface of the land above 32 degrees Fahrenheit, the freezing point of water. Exposed rock can warm above freezing even when the air temperature is below freezing because it absorbs radiation. Snow and ice atop the rock begin to melt, and water percolates into cracks and pore spaces. When the sun sets, temperatures drop rapidly. Water in small cavities in the rock freezes and expands, producing outward pressure that enlarges existing fractures and creates new ones. Every freeze-thaw cycle contributes to the ongoing disintegration or physical fragmentation of the rock.

Landscape Arch is a vast but delicate span.

Frost wedging is a physical process. Dissolution, on the other hand, is chemical. Calcium carbonate, the most important cementing agent in the various sandstones in Arches National Park, dissolves easily, and the chemistry of rain accelerates the dissolution. As rain falls, it interacts with carbon dioxide, the fourth most common gaseous component of the atmosphere, to form a weak solution of carbonic acid. So every time it rains, some of the calcium carbonate cementing the sandstones here dissolves and newly freed sand grains fall to the ground.

Balanced Rock is actually comprised of two different rock units, the underlying Dewey Bridge member and the overlying Slickrock member, both part of the Jurassic Entrada formation. It's easy to see where one member ends and the other begins. The Dewey Bridge member, which makes up the pedestal, has wavy beds with variable thicknesses and grain sizes, while the Slickrock member, which makes up the balanced caprock, is much more consistent in bedding thickness and grain size. Visible differences in the two units are a direct result of the environments in which they formed.

The Dewey Bridge member sits atop the massive, crossbedded Jurassic Navajo sandstone, discussed in vignette 8. The contact between the Navajo sandstone and the Dewey Bridge member is an unconformity, a gap in the stratigraphic record due to a period of either nondeposition

Balanced Rock, made of Entrada sandstone, marks the turnoff to the Windows and Double Arch.

or erosion. Immediately above the unconformity lie moderately thick and sturdy beds of fine sandstone. Farther up in the unit, the Dewey Bridge member changes, becoming deeper brown and composed of silty beds with irregular, wavy surfaces like those here at Balanced Rock. Rock units are reflections of ancient environments. Geologists think the Dewey Bridge member was deposited in tidal flats adjacent to a shallow sea to the west. Tidal flats are muddy lowlands crisscrossed by small channels through which tidal currents move. These flats are frequently flooded during high tides and exposed at low tides. Small changes in sea level can bury tidal flats in either beach dunes or deeper-water silts and muds. These environmental factors produce a mud-rich sandstone with a lot of internal variability.

The Slickrock member, which appears to have been deposited immediately after the Dewey Bridge member, is composed of fine sandstone in flat or crossbedded layers. Its sand grains are cemented by reddish orange iron oxide, which gives the unit its color, and calcite. In some areas of the park, varying quantities of iron oxide in this rock unit create bands of color within the Slickrock member. This unit formed in a beach dune environment, and the uniformity of both the dune sand and its cementation produces a strong unit that forms smooth cliffs.

The variability within the Dewey Bridge member makes it more vulnerable to frost wedging and dissolution than the Slickrock member, so weathering proceeds more rapidly on the underlying rock. At Balanced Rock, this dynamic has created a broad caprock of Slickrock sandstone resting upon a rapidly narrowing Dewey Bridge spire. At some point in the future, Balanced Rock will crash to the ground, leaving behind a short-lived pedestal.

This same process makes arches. To see how, proceed to stop 2. The Windows Trail takes you by a series of large arches at the interface between the Dewey Bridge and Slickrock members. In all of these arches, the contact between these two members is very obvious. Weathering has broken down the Dewey Bridge member more rapidly than the Slickrock member, undercutting the upper unit. Subsequently, the force of gravity worked in conjunction with weathering to cause fractures, then rockfalls, in the Slickrock member. Over time, this process creates expanding alcoves, deepening depressions in the rock face (this process is also discussed in vignette 9, about Zion National Park's Weeping Rock). If the bedrock stands in a fin, as it does here, the alcove will eventually break through to create a freestanding arch like the ones you see on this trail. Double Arch, made famous in the opening scenes of *Indiana Jones and the Last Crusade*, is an easy walk from the Windows parking lot. It, too, owes its formation to the juxtaposition of the weaker Dewey

The Windows are parallel arches that formed along the same plane of weakness in the Entrada formation.

Bridge member beneath the stronger Slickrock member. If these units were reversed, with the stronger rock below and the weaker rock above, no arches would form here. The upper unit would simply weather away more rapidly to expose the stronger bedrock unit below.

Stop 3 brings you to perhaps the most photographed and recognizable arch in the world, Delicate Arch (see photo, p. iv). The trail to Delicate Arch begins at Wolfe Ranch, a small cattle ranch that operated between 1888 and 1910. From Wolfe Ranch, the trail crosses Salt Wash on a small bridge. It winds up and around several promontories, then opens onto a broad surface composed of Slickrock sandstone. There is no trail here, but cairns and carved steps mark the way to the top. As you hike, look to your left, where a number of alcoves are forming along the cliff. This differential weathering indicates a zone of weakness in the otherwise strong rock wall. Also look at the major fractures beneath your feet; they're aligned with one another in an east-west direction, which is also the orientation of Cache Valley to the south. (We talk about the reasons for this in vignette 29.) The last part of the trail is a narrow path that has been blasted in a rock face with a steep drop to the left. One of the really neat things about this trail is that you never catch a glimpse of Delicate Arch until you turn the final corner, where it suddenly appears before you. The best time to walk this trail is late afternoon or early evening, when the dropping sun sets Delicate Arch aglow. The panorama is magical, with yellowish orange sandstone in the foreground set off against the majestic and often snow-covered La Sal Mountains on the horizon.

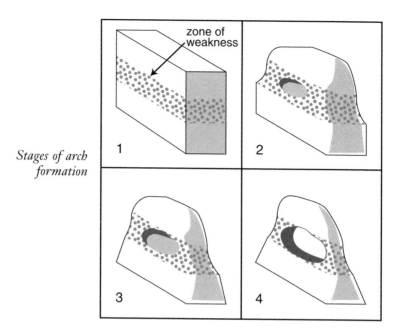

Stages of arch formation

Delicate Arch is much larger than it first seems; it's 50 feet high but rests on legs that are only 6 feet thick at their narrowest point. This arch, like the Windows arches, sits at the contact between two rock units, but here the underlying unit is the Slickrock member and the overlying whitish gray unit is the Moab Tongue member, the uppermost layer of the Entrada formation. Between the two members is a sharp unconformity marked by red siltstone and an indentation in the rock face resulting from this horizontal plane of weakness. The contact between the two formations is easy to identify. The north leg of Delicate Arch shows a pronounced indentation about halfway up where the leg is thinnest. This is the location of the contact and the point where weathering occurs most rapidly. The overlying Moab Tongue member formed in a beach dune environment very similar to that in which the Slickrock member formed, but it's lighter in color due to its predominantly calcareous cement. Like the Slickrock member, it forms resistant cliffs and slopes. But differential weathering doesn't require much to gain a foothold; the contact that separates the two units here is enough to allow frost wedging, dissolution, and other weathering processes to advance into the rock face, creating alcoves and finally arches. Once again, as the arch grows, it expands upward into the overlying unit—here, the Moab Tongue member—as gravity pulls rock down and away from the underside of the arch. Delicate Arch is all that remains of a fin that once stood here, aligned in the same direction as the fractures you strode over on the bedrock trail.

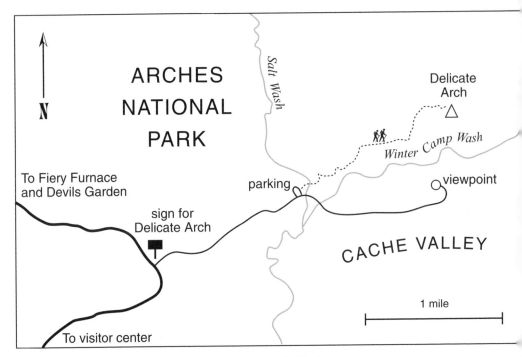

Delicate Arch Trail

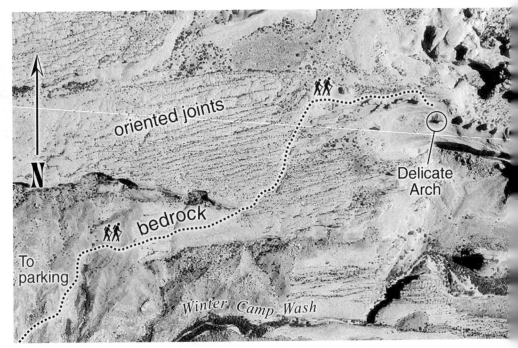

Aerial photo of the trail to Delicate Arch

Other mechanisms have created arches in fins elsewhere in the park. In the Fiery Furnace and Devils Garden (discussed in vignette 29), sand has filled in the space between neighboring fins. This sand holds moisture from rainfall against adjacent rock faces. This concentrates carbonic acid, and hence dissolution, at a particular height in the rock; in this scenario, weathering creates a zone of weakness in rock where none may have existed before. Once depressions form, they create more surface area vulnerable to attack by both dissolution and frost wedging, eventually creating arches as deepening alcoves on both sides of a fin join in the center. In other cases, fractures or weakly cemented sections within the rock promote arch formation. But the degree of fracturing is important. If the rock is too highly fractured, it will be too weak to stand against the ever-present force of gravity, preventing arch formation.

Arches are found throughout southern Utah, but nowhere in such abundance as at Arches National Park. They can be large or small, delicate or sturdy. Some are symmetrical, but others, like Double Arch, are contorted and unique. One thing these graceful structures do have in common is that all are born of an exquisite balance between strength and weakness. Stroll some of the park's trails and examine the rock to identify the sources of weakness that promote differential weathering as you appreciate the strength that allows the arches to stand tall. If you're interested in seeing other unique features associated with the Entrada sandstone, visit Goblin Valley, the subject of vignette 21.

Alcoves north of the trail to Delicate Arch mark the onset of weathering and erosion that may form future arches.

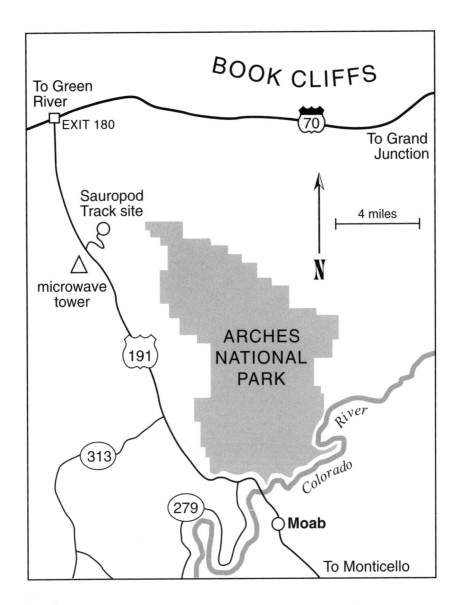

GETTING THERE

From exit 180 on I-70, drive 8 miles south on U.S. 191. Or, from Moab, drive 23 miles north on U.S. 191. At 0.75 mile north of milepost 148, turn east on a graded dirt road signed for Copper Ridge Dinosaur Tracks. Follow signs to the parking area and take the short trail (about 100 feet) to the tracks.

31 IN THE FOOTSTEPS OF GIANTS
Copper Ridge Dinosaur Tracks

At Copper Ridge, preserved prints of several long-dead animals lie atop a low ridge of sandstone. The tracks are exposed on a bedding plane (a division between two rock layers) that was once a muddy flat across which these animals walked. When prints appear in sets or collections, as they do here, they're called trackways. Recognizable three-toed prints move back and forth across the cleared rock surface, and at the western edge of the trackway, there are also large depressions that look as though they were created by weathering—that is, until you look more closely. Most of these depressions are round and measure several feet across and 6 to 12 inches deep. Interspersed among the rounded pits are equally deep crescent-shaped depressions. Those deep pits are why you're here, and they're also what make this trackway special.

The pits are actually footprints of a four-footed sauropod, one of the giants of Mesozoic time. This great beast, weighing tens or perhaps hundreds of thousands of pounds, strode onto this surface 150 million years ago. The rock beneath your feet belongs to the Morrison formation, well known for its treasure trove of fossil dinosaur bones, teeth, and claws. But dinosaur trackways are rare in the Morrison—only about twenty of them have been discovered in this formation in ten states. Most of those sites have only a few scattered three-toed prints. The site at Copper Ridge is important because of the diversity of prints and also because of the presence of the sauropod tracks, which are not nearly as common as other types of dinosaur prints. This was the first sauropod trackway discovered in Utah, and it wasn't found until 1989.

All dinosaurs belonged to one of two groups, or orders: Ornithischia and Saurischia. They are classified based on the geometry of their hip bones. *Ornithischian* means "bird hipped," a reference to the similarity between the structure of this order's hips and that of birds; *saurischian* means "lizard hipped." Ornithischians comprised 45 percent of all known dinosaurs, with saurischians making up the other 55 percent. All ornithischians were herbivores, while saurischians included both carnivores and herbivores. Paleontologists further subdivide Saurischia into two subgroups, Theropoda and Sauropoda. Thanks in part to Hollywood movies, the best-known theropods are probably *Tyrannosaurus*, a 20-foot-tall hunter and scavenger who weighed as much as a modern elephant, and *Velociraptor*, a 6-foot-tall predator of incredible speed. Theropods had sharp teeth and were bipedal—that is, they ran and walked on two legs. They bore little resemblance to their relatives, the sauropods, who were quadrupedal, walking on four legs.

Sauropods were the biggest, heaviest land animals in the history of the earth. When sauropod bones

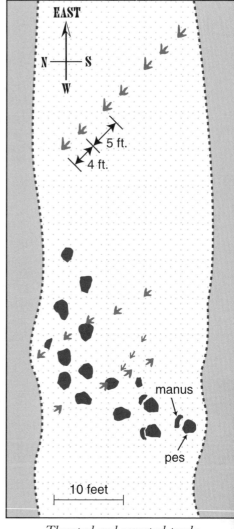

Theropod and sauropod tracks at Copper Ridge

were first discovered, their immense size and porous bone structure, similar to whale bones, earned the animals the name "whale lizards." Four stout legs supported a massive body. At one end was a long tail, and at the other, an equally long neck and a small head. Adult sauropods varied in size from the 30-foot-long *Amargasaurus* to the 125-foot-long *Seismosaurus*, the longest, heaviest creature that ever walked the planet. Adult seismosaurs probably weighed well over 100 tons! Although quadrupedal, some sauropods probably stood on their hind legs briefly while foraging for leaves.

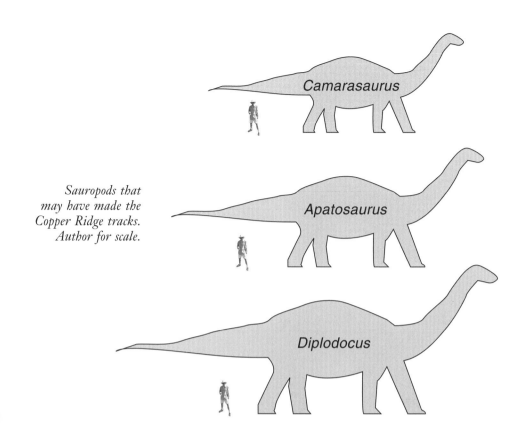

Sauropods that may have made the Copper Ridge tracks. Author for scale.

Camarasaurus

Apatosaurus

Diplodocus

The site is small, but sauropod tracks are relatively rare in the geologic record.

A visitor has marked three-toed theropod tracks with gravel rings.

At this small site, you can see five individual trackways, which were made by three large theropods, one small theropod, and one sauropod. If possible, look at the prints in the early morning or late afternoon, when strong shadows help delineate the impressions. At the west end of the site, things are complicated. The very large, amorphous depressions, which are in fact sauropod prints, head to the northeast, then turn eastward at the northern margin of cleared rock. The sauropod trackway is quite narrow, resembling the tracks of modern-day elephants, which walk by placing one foot almost exactly in front of the other. Two large sets of theropod tracks are superimposed on the sauropod trackway; one set of prints heads to the southeast, while the other set leads to the northwest. A final set of small theropod prints points southeast. At the east end of the site, large, three-toed theropod prints head off in a southeasterly direction.

Sauropods had a large, fleshy cushion behind each toe on their back feet. This allows us to differentiate tracks made by their front and back feet. The large, circular depressions mentioned before are from

A sauropod print lacks the clear indication of toes seen in theropod prints.

their back feet. Paleontologists call these pes prints; *pes*, pronounced PEHZ, means "hind foot" in Latin. There are also smaller, less symmetrical depressions that are both wide and thin. These are manus (MAN-uss) prints made by the sauropod's front feet; *manus* means "forefoot" or "hand." Based on evidence from trackways and fossils, sauropods' front feet had vertical foot bones. The way they walked would be similar to you crawling with the weight of your upper body on your fingertips instead of the palms of your hands. Pes prints obscure many of the manus prints at this location, but some are still visible.

Fossils have shown us that dinosaur bone and muscle structure were similar to those of existing organisms. Based on this, we can tell quite a bit about a dinosaur's size from its tracks. We know that for specific dinosaurs, footprint length is related to the distance from the ground to the hip, and hip height in turn correlates with body length and mass. In most sauropods, height at the hip equals roughly six times the pes footprint length. Since the Copper Ridge pes depressions are about 2 feet long, the hip height of the sauropod that ambled here was about 12 feet—2 feet higher than a basketball hoop. That's big! Because these are just footprints, and ancient ones at that, we may never be able to determine exactly which type of sauropod made these tracks. Scientists

Deep three-toed print

have suggested it was a *Camarasaurus*, *Apatosaurus*, or *Diplodocus*, and fossil bones from all three of these dinosaurs have been found in the Morrison formation. Though we may never know which sauropod made these tracks, our formula tells us this massive creature was about 70 feet long from tail to nose.

We can also estimate body size from the theropod prints. Theropods had hip heights equal to about five times their footprint length. The larger prints are just over 12 inches long, so those were made by theropods with hip heights of about 5 feet, which is consistent with *Allosaurus*, the carnivore most commonly represented by fossil bones in the Morrison formation. These creatures were about 15 feet tall and weighed between 2,000 and 4,000 pounds.

Stride length is also telling. Look at the easternmost theropod trackway, where it's easy to measure stride length—the distance from one right footprint to the next right footprint (a complete stride is actually two paces). In this case, stride length is about 9 feet. In modern bipeds, the ratio of stride length to hip height can be used to determine if the track maker was walking, trotting, or running. Hip height, of course, cannot change, but stride length can. When an animal walks faster, stride length increases slightly, and when an animal begins to run, stride length increases dramatically. The ratio of stride length to hip height for this track is 1.8, so, based on ratios for modern bipeds, we can surmise that the theropod was walking. But it wasn't walking smoothly. Close inspection shows alternating short and long paces. A 4-foot-long pace is followed by a 5-foot-long pace in a consistent pattern across the outcrop. It has been suggested that this animal was limping because of a recent injury. Perhaps the limp resulted from a broken bone that mended poorly. Or maybe the animal was burdened. Could this dinosaur have been walking unevenly because of the weight of a bloody chunk of flesh hanging from its massive jaws?

Based on these tracks preserved in Morrison formation sandstone, we know that several dinosaurs, including four theropods and a rarer sauropod, walked across this muddy plain at approximately the same time. We'll never know exactly what caused the uneven gait of one of the theropods as it strode across this muddy surface 150 million years ago, just as we'll never know what caused the sauropod to pause, then turn to the east. We can, however, tell something about the size and behavior of these creatures from simple mathematical relationships between footprint length, hip height, and stride length. The clues present at the Copper Ridge trackway allow us to more fully imagine the scenario that played out here—just one small episode in the larger saga of the long-extinct dinosaurs, magnificent rulers of Mesozoic earth.

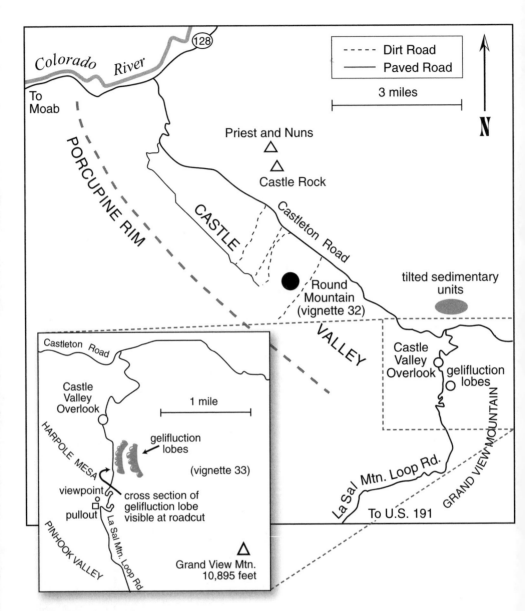

GETTING THERE

From exit 180 on I-70, take U.S. 191 south 27 miles to the turnoff to Utah 128. From Moab, drive 3 miles north on U.S. 191 to that same turnoff. Take Utah 128 east 16 miles to Castleton Road (there's a sign here for La Sal Mountain Loop Road), then go 11 miles southeast on Castleton Road to La Sal Mountain Loop Road. Turn right and drive 1.8 miles to the Castle Valley Overlook on the right.

32 INTRUDERS IN A SEDIMENTARY DOMAIN
Round Mountain and the La Sal Laccoliths

As you drive Utah 128 along the Colorado River, you see fantastic outcrops of flat-lying red Mesozoic sandstone and shale, so characteristic of the Colorado Plateau. When you turn south onto Castleton Road, though, you begin to see something quite different—the La Sal Mountains, which rise to a maximum elevation of 12,721 feet on Mount Peale. The La Sal Mountains, in the Manti–La Sal National Forest, are made up of igneous rock that intruded country rock (another name for preexisting bedrock). In this case, the country rock is sedimentary. This intrusion formed below the surface, but tens of millions of years of erosion of overlying material has exposed it.

Castleton Road takes you southeast into Castle Valley. In the very center of the valley stands a dark, brooding hulk of rock, which, together with the bulk of the La Sals, is the subject of this vignette. Most new visitors mistakenly assume that this is Castle Rock, after which the valley was named, but the dark mass is called Round Mountain. Castle Rock is the sandstone spire on the northeast side of the valley close to another sandstone landform, the Priest and Nuns. To the southwest, Castle Valley is bounded by Porcupine Rim, an immense sandstone cliff; rock layers there angle up toward the center of Castle Valley. This is evidence of a salt anticline that domed overlying rock upward before its salt core dissolved and collapsed to form Castle Valley (we discuss this process in greater detail in vignette 29). The rocks at the edge of the valley still retain their domed orientation.

As you follow the directions in "Getting There," pay attention to the layered sedimentary rocks on the left side of the road as you approach

Round Mountain dominates Castle Valley.

the turnoff for La Sal Mountain Loop Road. They lie at a steep angle against the northern flank of the La Sals, as if pushed aside by the rising mountains. The slope of these layers is unrelated to the salt anticline that domed the rock along Porcupine Rim. Sedimentary units on the southwestern flank of the La Sals also stand at a vertical angle, attesting to powerful forces that accompanied emplacement of rock here.

Roadcuts are valuable sources of information for geologists. They allow us to see unweathered rock surfaces, without the patina that develops with time and exposure. In vignette 27, on Newspaper Rock, we talk about desert varnish, the manganese-rich filmy coating that forms on exposed rock in arid and semiarid environments. It might surprise you to know that the rock that makes up Round Mountain, for example, isn't black; it's light gray. However, the surface is covered with desert varnish. Lichens also flourish on some rock surfaces, affecting their color, and interactions between atmospheric gases and rock often alter minerals and change surface colors and textures. Rock identification relies on pristine faces that clearly show the rock's mineral crystals. (Minerals are the fundamental building blocks of rocks, and all minerals are crystals.) Roadcuts provide us with recent exposures that haven't had time to change color or mineral composition. As you drive La Sal Mountain Loop Road to the overlook, you'll pass several roadcuts. Some are red

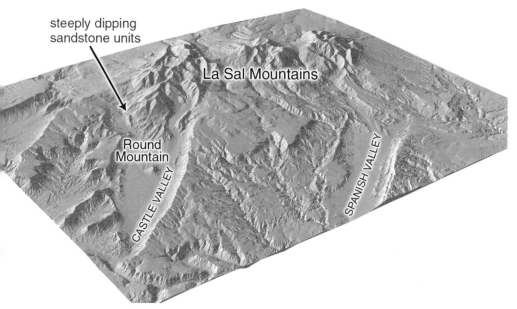

steeply dipping
sandstone units

La Sal Mountains

Round
Mountain

CASTLE VALLEY

SPANISH VALLEY

Three-dimensional image of the La Sal Mountains

sandstone; others are light gray. Safely stop in a pullout near one of the gray exposures to take a closer look.

Color tells us something about a rock's chemistry: minerals rich in silica tend to be clear, white, or pink, while minerals rich in iron and magnesium tend to be black. The gray color you see here reflects minerals with roughly equal quantities of silica on the one hand and iron and magnesium on the other. If you look closely at a hunk of this gray rock, you can see oblong and chunky black and white fragments, all less than 0.35 inch long, dispersed within a light gray background material called the matrix. A magnifying lens would reveal tiny, discrete mineral grains within the salt-and-pepper texture of the matrix. A great deal of information can be gleaned from this simple textural analysis. Mineral crystals grow as molten rock cools and crystallizes; the size of the crystals reflects cooling time. If magma remains far underground and cools very slowly, large crystals form in a mass of rock called a pluton. The crystallization of lava is on the opposite end of the spectrum. Molten rock on the earth's surface cools very rapidly and forms microscopic crystals. In fact, lava may cool so rapidly that no minerals form at all. Instead, the lava becomes obsidian, natural glass that totally lacks the regular arrangement of elements found in crystalline mineral structures (we discuss the special characteristics of obsidian in vignette 14).

In the roadcut before you, the rock with larger minerals set into a matrix of finer minerals is called a porphyry. A porphyry is a rock with two histories. The large minerals here, hornblende (black) and plagioclase (white), obviously formed during slow cooling at depth, but the fine matrix resulted from rapid cooling near the surface. How did these two types of crystals come to be intermixed? Temperatures within the earth increase with depth. So when magma moves slowly upward through the crust, the slow decrease in temperature results in the formation of sizable hornblende and plagioclase crystals. The Colorado Plateau is a thick mass of crustal rock that presents a formidable obstacle to upward movement. On the margins of the Colorado Plateau, resistance is weaker; magma was able to breach the surface, resulting in the Marysvale volcanic field to the west and the San Juan volcanic field to the east. Geologists think the La Sal magma body stopped about 1 to 3 miles below the surface, where cool near-surface temperatures caused rapid crystallization of the fine matrix. So what do we call this rock? An intermediate matrix with inset hornblende and plagioclase crystals is called a hornblende plagioclase trachyte. You might also notice some coin-size dark gray or black amorphous chunks that stand out against the lighter-colored trachyte. As magma moves upward, it incorporates pieces of the rock it pushes through. These pieces are called xenoliths (*xenolith* means "foreign rock"), and here they appear as the darker blobs within the trachyte.

Now proceed to the Castle Valley Overlook, which sits on the western flank of 10,895-foot Grand View Mountain. To the west, Round Mountain dominates the view of Castle Valley. Grand View Mountain

A close look at the La Sal Mountain trachyte and a dark xenolith, a piece of country rock carried upward with the moving magma

and Round Mountain are both part of the intrusion that created the La Sal Mountains. Three large clusters of igneous rock make up the northern, central, and southern La Sal Mountains; you're standing on the western flank of the southernmost La Sal cluster. Radiometric dating of the La Sal Mountain trachyte yields ages between 28 and 25 million years old, so crystallization occurred in late Oligocene time. Geologists think that preexisting faults and fractures, some associated with movement of Paradox formation salt, created weak zones that allowed magma to push through almost to the surface. Magma moved upward to form laccoliths, large mushroom-shaped mounds wherein magma pushes in between two rock layers and domes the overlying rock upward. Each of the three massive bodies of igneous rock that make up the La Sals is a laccolith. Subsequent erosion of overlying layers of sedimentary rock has revealed these impressive mountains. Steeply angled rock layers on the mountains' flanks are evidence of the original doming.

Round Mountain is a satellite intrusion. Underground it's connected to the same system that created the larger mass of mountains around it. A well-developed layer of desert varnish has darkened the surface, but underneath that patina lies the same hornblende plagioclase trachyte that makes up the rest of the La Sal Mountains. Round Mountain displays some other interesting characteristics. You may want to head down into Castle Valley and walk or drive one of the gravel roads that lead from Castleton Road to the base of Round Mountain. If you do, pay attention to the red sandstone as you approach the base of Round Mountain. Stop and bang some pieces together. Far away from the Round Mountain pluton, sandstone pieces thud against one another, but near the base, they

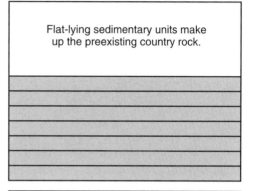

Flat-lying sedimentary units make up the preexisting country rock.

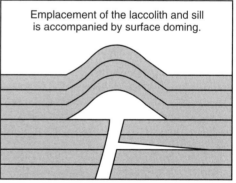

Emplacement of the laccolith and sill is accompanied by surface doming.

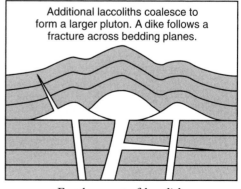

Additional laccoliths coalesce to form a larger pluton. A dike follows a fracture across bedding planes.

Emplacement of laccoliths

clang like fire-hardened tiles. That's an apt analogy because magma, at temperatures often exceeding several thousand degrees Fahrenheit, bakes country rock to form a metamorphic aureole, or rim, that completely surrounds it. Harder metamorphic rock rings like a bell—in this case, a bell made of metamorphosed sandstone. Remnants of the baked sandstone aureole also appear at the top of Round Mountain; these are known as roof pendants because of their position atop the laccolith. The roof pendants tell us that the magma body stopped beneath sandstone bedrock here and never reached the surface.

Plutonic mountain ranges can be found throughout the western United States. They vary in size from the vast Sierra Nevada of California to much smaller unnamed plutons that are barely hills. The La

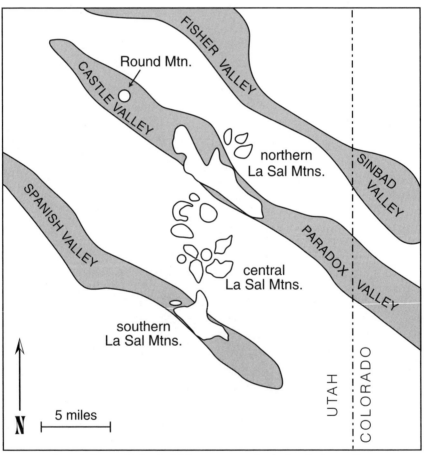

Laccoliths in the La Sal Mountains and salt dissolution valleys

Sals are not the only plutonic range in southern Utah. There are two others of similar age and structure nearby: the Abajo Mountains to the south and the mysterious Henry Mountains to the southwest (the last mountain range in the Lower 48 to be explored). After intrusion of the La Sal, Abajo, and Henry Mountains, magmatic activity within the Colorado Plateau ceased for a period of 15 million years. Erosion took over, and the Colorado River and its tributaries carried away miles of rock that once overlay the three laccoliths. Today these mountains give us a glimpse into the behavior of molten rock below the earth's surface, as well as stunning alpine landscapes that contrast sharply with a modern desert dominated by sedimentary rock.

Metamorphosed sandstone surrounds Round Mountain. Unlike sandstones farther away, this metamorphic rock rings like a bell when struck with a hammer. Lens cap for scale.

Periglacial flow lobes on the western flank of Grand View Mountain

GETTING THERE

From exit 180 on I-70, take U.S. 191 south 27 miles to the turnoff to Utah 128 (see map on p. 252). From Moab, drive 3 miles north on U.S. 191 to that same turnoff. Take Utah 128 east 16 miles to Castleton Road (there's a sign here for La Sal Mountain Loop Road), then go 11 miles southeast on Castleton Road to La Sal Mountain Loop Road. Turn right and drive 3 miles to a pullout at the base of a low hill on the right side of the road. The crest of this hill offers a good view of the mountain slope to the northeast.

33 PERMAFROST AND FLOWING EARTH
Periglacial Features in the La Sal Mountains

The ice ages, a general name assigned to Pleistocene time, saw periodic growth and decline of continental ice sheets and valley glaciers for a period of almost 2 million years. During this time, the La Sal Mountains behaved like most very high mountain ranges in the southwestern United States. During cooler climate cycles, they hosted glaciers, which then melted in the intervening warm periods. We talked about glaciers in southern Utah in vignette 15, about Fish Lake Valley. But this vignette isn't about glaciers; rather, it's about periglacial processes. *Periglacial* literally means "near glaciers." It denotes landforms that develop where permafrost, permanently frozen ground, is present.

Look northeast at the broad, largely unvegetated slope just above La Sal Mountain Loop Road. This is part of the western flank of Grand View Mountain, the 10,895-foot mountain to the east. In the morning or late afternoon, you can clearly see large semicircular ripples on the hillside; they are a bit more difficult to see in the harsh midday sun. The ripples, referred to as lobes, are roughly parallel to one another, and the hillside itself looks as if it once flowed like hot wax. Use of the word *flow* is accurate, although the analogy of hot wax is not. Flow here was controlled by something much, much colder.

Permafrost is permanently frozen ground in which pore spaces between soil particles and rock fragments are completely filled with ice crystals. Where do we find permafrost? Well, underground, of course, and in geographic regions that are cold even in summer. In the ice ages, permafrost was a common feature in mountain ranges, occurring in high-elevation areas between glaciated valleys. However, if you had visited one

A roadcut reveals the broken rock and soil that underlies the gelifluction lobes.

of these ranges 20,000 years ago, at the height of the last glacial stage, and poked a stick into the ground in mid-July, you would have discovered a squishy, muddy slope, not rock-hard ice. Though permafrost was present at that time, it wasn't located right at the surface.

In winter, soils underlain by permafrost freeze completely. In summer, though, intense solar radiation thaws the upper soil layer. Since heat from the sun doesn't reach deeper soil layers, they remain frozen. The upper portion of soil that freezes in winter and thaws in summer is called the active layer. When the active layer warms, its ice melts and liquid water fills its pore spaces. In today's warmer climate, this water would simply percolate downward into the deeper soil, but in periglacial environments, permafrost prevents infiltration. Since pore spaces filled with ice are impermeable, water in the active layer has nowhere to go. This creates a saturated layer on top of the permafrost.

Saturation presents an interesting problem. Water in pore spaces exerts outward pressure that pushes soil grains apart. Friction between grains is what holds soil on sloped surfaces up against the pull of gravity. If grains are pushed apart, friction decreases and gravity gains the upper hand. The saturated slope begins to flow. In periglacial regions,

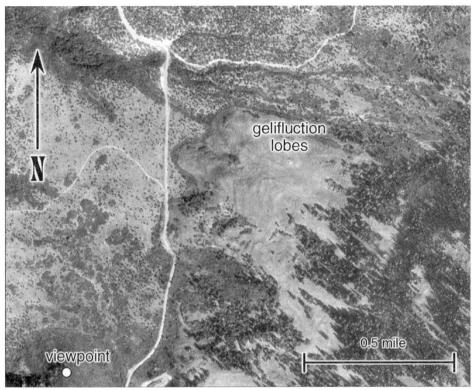

Aerial photo of flow structures

this movement is intermittent and slow. Brief summers of downslope movement are followed by long winters of ice-hard stasis. This process is called *gelifluction*, which means "frozen flow," despite the fact that flow occurs in warm, not cold, seasons. *Gel* is a good description of the saturated active layer, which slowly oozes downslope as a gooey mass—not quite liquid, yet not quite solid. Look at the slope to the northeast to see this activity captured in time. This is essentially how the slope looked when the last permafrost melted and the process of gelifluction ceased.

The La Sal Mountains have other periglacial features. In the high valleys here, once occupied by true glaciers made of ice, there's also evidence of rock glaciers, amalgams of rock cemented together by ice. Rock glaciers are found, for example, beneath both Mount Peale and Mount Tukuhnikivatz in the central La Sal Mountains. You can hike in to get a close look at these features. The trailhead is 11 miles south on La Sal Mountain Loop Road, then 6 miles east on Geyser Pass Road. Turn south (right) on Forest Road 241 (some maps show this as Forest Road 141) and drive 1 mile to the trailhead. The trail, 2 miles long, follows an old roadbed. Along the way, you'll be rewarded with tremendous views of high peaks, glacial lakes, alpine meadows, and landscapes shaped by both glacial and periglacial processes.

Another option is simply to look at the aerial photograph on the next page. When climate slowly warms after glacial stages, glaciers melt completely, but that doesn't mean all movement suddenly stops. Here in the La Sals, the postglacial climate remained cool enough to produce more abundant winter snow than occurs today, and rock debris falling from cliff walls covered that heavy snowpack. Compression from the accumulated weight of rock and snow produced ice that glued the rocks together. This mass of ice and rock began to flow, though not so smoothly, so quickly, or so far as true glacial ice flows. With additional warming, even this process ceased. These rock glaciers are recognizable today by their elongate shapes and rippled upper surfaces, which, like the ripples on the Grand View Mountain slope we've been discussing, indicate past flow.

How can we be sure that the slope to the northeast isn't still flowing? The answer lies in the presence of scattered trees, some rather large, on the hillside. If this region were still flowing, any trees growing on this slope would topple in a downslope direction. No tree in this already stressful alpine environment could grow quickly enough to offset movement in the active layer. So the upright trees here tell us this slope is at rest—for the time being. Some scientists believe that our planet is in the latter stages of an interglacial period. If they're correct, climate may cool over the next several thousand years and drive us into a glacial stage once again. Of course, there's some question about whether our future

An extinct rock glacier pours out of a high valley below Mount Tukuhnikivatz.

will be cooler or warmer, given the greenhouse effect and the potential for global warming. But if we do enter a new glacial stage, ice will once again dominate the high valleys here and permafrost will develop in the deeper soil layers. A rejuvenated active layer will ooze downslope once more, and glaciers will resume their work of shaping the La Sals.

Permanent chutes form where snow avalanches continuously clean slopes of vegetation.

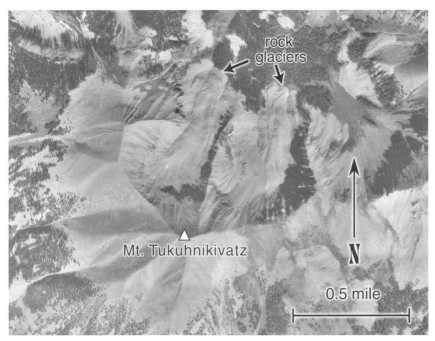

Air photo of rock glaciers beneath Mount Tukuhnikivatz

Canyon tree frogs near Weeping Rock, Zion National Park. The frogs make a whirring sound reminiscent of a distant machine gun.

GLOSSARY

aa. Viscous lava that forms blocky, angular flow surfaces.

alluvium. Unconsolidated sediment deposited by running water.

amalgam. A mixture of mercury and valuable metals resulting from the separation of mineral and waste.

anthropomorph. A depiction of an object or animal with human characteristics or features (as in rock art).

anticline. An up-arching fold of rock strata.

anticyclone. A region of atmospheric high pressure.

ash flow. A hot, rapidly flowing mixture of ash, volcanic gases, and pumice created during explosive volcanic eruptions.

ash-flow tuff. A volcanic rock composed of material from an ash flow.

asphaltite. Carbon-rich deposit that hosts uranium and vanadium.

assay. A test to determine economic viability of a particular ore.

asthenosphere. The part of the upper mantle that lies immediately below the lithosphere. Convective motion of the viscous material within the asthenosphere drives plate motion above.

basalt. A fine-grained, dark, extrusive igneous rock containing calcic plagioclase, pyroxene, and olivine, with little silica.

basement. Old rocks, generally igneous and/or metamorphic, that lie beneath the younger rock and sediment in an area.

bedding. The layered structure of sedimentary rocks.

bedrock. Solid rock, either exposed or lying beneath a mantle of unconsolidated material.

braided stream. A stream pattern that forms when the sediment load exceeds the transport ability of a stream. Material deposited as channel bars causes the stream to continually divide and rejoin.

butte. A steep-sided, flat-topped, isolated bedrock landform.

caldera. A huge bowl-shaped depression formed by the collapse of a volcanic cone following a large eruption.

caliche. A subsurface soil layer of hardened calcium carbonate that develops in arid and semiarid climates.

chemical weathering. The chemical alteration of rock at the earth's surface.

cinder cone. A conical accumulation of cinders, or gravel-sized pieces of basalt.

cirque. A bowl-shaped, glacially carved depression at the head of a mountain valley.

clast. An individual fragment of rock.

conglomerate. A sedimentary rock composed of rounded pebbles and gravel from preexisting rock.

contact. The point at which two formations touch.

country rock. Preexisting rock through which magma moves as it rises toward the earth's surface.

crossbeds, also **crossbedding.** Inclined sedimentary layers that indicate wind or water direction during deposition.

crustal extension. The pulling apart of the earth's crust due to upwelling in the asthenosphere.

daughter product. An element that results from radioactive decay of a parent isotope.

deformation. Physical alteration of rock due to an applied stress.

desert varnish. A patina of iron, manganese, and clay that builds up on exposed rock surfaces in arid environments. Also called **rock varnish**.

diapir. An upward intrusion that pierces an overlying rock structure.

differential weathering. The uneven effects of weathering on variable rock types, causing them to break down at different rates.

dike. A tabular intrusion of igneous rock that cuts across the rock into which it has intruded.

dip. The angle that any planar geologic surface, such as a bedding plane or fault plane, makes with the horizontal. It is measured perpendicular to strike. See **strike**.

downwarp. A downward bend in the earth's crust.

drag. See **fault drag**.

epeirogeny. Uplift of large landmasses without extensive internal deformation.

epicenter. The point on the earth's surface directly above the point of rupture, or focus, of an earthquake.

erosion. The removal and transport of rock and sediment by water, wind, and ice.

evaporites. Minerals that form during evaporation of a standing body of water.

extrusive rock. Igneous rock that forms from lava.

fault. A fracture in rock along which relative motion has occurred.

fault block. A block of rock bounded by faults on at least two sides.

fault drag. The bending up or down of layers of rock as a result of movement along a fault. Also known as **drag**.

fault line. The intersection of a fault plane with the surface of the earth; the surface expression of a fault.

fault plane. A fault surface without notable curvature.

fault scarp. A steep slope or cliff formed by vertical movement along a fault.

feldspar. The most widespread group of rock-forming minerals, containing primarily silica, aluminum, and oxygen combined with calcium, sodium, and/or potassium.

fin. A tall, narrow ridge of rock created by fracturing and subsequent weathering.

firn. Coarse, granular ice that forms from compression of snow. Further compression creates true glacial ice.

focus. The point of first rupture during an earthquake. The focus lies beneath the earth's surface and underlies the epicenter, which lies on the earth's surface.

fold. A bend in rock strata.

footwall. The lower or underlying side of an inclined fault plane. You could stand on the footwall (hence the name) if the overlying fault block were removed.

formation. A mappable, distinct rock unit deposited under a uniform set of conditions. Formations may be subdivided into members.

fossil. Remains of an organism preserved in rock.

frost wedging. The process of mechanical disintegration of rock due to pressure exerted by water in cracks or joints.

gelifluction. The downslope movement of debris induced by seasonal thawing and freezing in an area underlain by permafrost.

geomorphic province. A region of the earth's surface identified by its characteristic landscape.

glacier. A large mass of flowing ice.

graben. A fault-bounded block of the earth's crust that has dropped down relative to neighboring blocks.

granite. A coarse-grained, intrusive igneous rock composed primarily of quartz and alkali feldspar, with lesser amounts of plagioclase and mica.

ground moraine. A horizontal layer of rock debris deposited directly onto the valley floor during glacier retreat.

gypsum. A sulfate mineral known to develop in hydrothermal deposits and during evaporation of surface water.

half-life. The time required for one-half of a radioactive isotope to decay into a stable daughter product.

hanging wall. The upper or overhanging side of an inclined fault plane.

hoodoo. A prominent spire that forms from rapid erosion of a weak substrate.

hornblende. A common, dark, rock-forming silicate mineral in the amphibole group that often forms elongated crystals.

hydrothermal. Related to hot water beneath the earth's surface.

igneous rock. Rock formed from the cooling and crystallization of lava or magma.

ignimbrite. Rock formed by the consolidation of hot ash flows from explosive volcanic eruptions. Ignimbrites contain pumice fragments and may or may not be welded.

intermittent stream. A stream that flows only part of the year, either during a rainy period or shortly thereafter.

intrusive rock. A rock body that forms from magma beneath the earth's surface.

inverted topography. A configuration in which older rock units sit above younger units.

isostasy. The state of gravitational equilibrium between the viscous asthenosphere and the brittle lithosphere.

joint. A fracture in a rock with no displacement or relative motion.

laccolith. A large mushroom-shaped intrusive igneous rock.

landslide. A general term for gravity-driven movement of rock and debris downslope.

lateral moraine. Rock debris deposited by a glacier on the sides of a valley.

lava. Molten rock material on the earth's surface.

lee. Refers to the side of a landform that is sheltered from the wind.

lens. A deposit that is thick in the middle and thins toward the edges.

lichen. Algal and fungal organisms that coexist and take nourishment directly from rock.

limestone. A sedimentary rock composed of calcium carbonate.

lithic. A fragment of previously formed rock, or rock that contains such fragments.

lithification. The process of compaction and cementation that transforms sediment into sedimentary rock.

lithosphere. The brittle upper portion of the earth, consisting of the crust and uppermost mantle.

magma. Molten rock beneath the earth's surface.

member. A subdivision of a defined rock formation.

metamorphic rock. Rock changed by heat and/or pressure to become distinct in texture or composition from its parent rock.

mesa. A flat-topped landform with steep, sloping sides that generally covers a large area.

meteor. A body of rock in space that is neither a planet, moon, nor asteroid. So-called "shooting stars" are meteors that burn up due to intense friction between atmospheric gases and rock that moves at thousands of miles per hour.

meteorite. A meteor that strikes the earth's surface.

moraine. An accumulation of unsorted, unstratified debris deposited by a glacier.

nodule. A small lump or knot.

normal fault. An inclined fault in which the hanging wall moves downward relative to the footwall as a result of extension of the crust.

obsidian. Dark volcanic glass that forms when lava cools so rapidly that no crystals can form.

offset. The distance two points on opposite sides of a fault move in relation to one another; the amount of movement along a fault in a given time period.

ore. An economically viable mineral deposit. Most commonly, deposits of metallic minerals such as gold, silver, and copper.

orogeny. A period of folding, faulting, and thrusting during which mountain ranges form.

oxidation. The combination of an element or compound with oxygen.

pahoehoe. Fluid lava that cools to form a ropey upper surface.

paleoichnology. The study of footprints and other traces of ancient organisms.

permafrost. Permanently frozen soil.

permeability. The ability of rock or sediment to transmit a fluid. A function of the interconnectedness of pore spaces in rock.

petrification. Replacement of organic material such as wood or bone by minerals.

petroglyph. A drawing chipped or carved into a rock surface.

physiographic. Pertaining to the earth's surface features and landscapes.

pictograph. A painting on a rock surface.

plagioclase. Common rock-forming minerals of the feldspar group that contain sodium and/or calcium.

plate. A defined area of the lithosphere that moves as a unit.

playa. An unvegetated flat surface left behind during the evaporation of a lake. Often contains a variety of salts.

pluton. A body of igneous rock that crystallizes beneath the earth's surface.

pluvial lake. A lake that varies in size with climate change.

pore space. A space between rock grains that is filled with air or water.

porosity. The total volume of rock or sediment that is made up of voids or pore spaces.

porphyry. An igneous rock in which larger crystals that grew at depth early in the cooling process sit within a matrix of much finer crystals that grew during a later episode of more rapid cooling.

pumice. Lightweight volcanic glass containing vesicles from gas that escaped from lava during cooling.

quartz. A common rock-forming mineral composed solely of silica and oxygen.

radiocarbon dating. A method of establishing the age of dead organic material based on the steady decay of carbon-14 to nitrogen-14.

radiometric dating. A method of determining absolute ages for rocks based on the decay of radioactive isotopes.

radium. A radioactive element found in uranium ores such as pitchblende.

radon. A radioactive gaseous element that forms as radium disintegrates. Colorless and odorless.

recessional moraine. An accumulation of unsorted, unstratified debris deposited by a receding glacier.

regressive sequence. A sequence of rocks deposited as sea level retreats.

reverse fault. A fault in which the hanging wall has moved up relative to the footwall due to compression.

rhizoconcretion. A hardened sediment cylinder around a plant root.

roadcut. Mostly unweathered and previously buried rock that is exposed by road construction.

rockfall. Free fall of a freshly detached rock mass from a cliff.

saltation. Bouncing movement of sediment particles along the desert floor or a stream channel.

sedimentary rock. Rock made up of weathered remains of preexisting rock.

shale. A sedimentary rock consisting mainly of clay minerals, the laminate structure of which causes it to break along parallel planes.

shear. Deformation resulting from forces that cause two rock bodies to slide past one another along a plane or zone.

silica. Common term for the compound silicon dioxide.

slickensides. Fine grooves resulting from rock grinding against rock during movement along a fault plane.

slickrock. A smooth, weathered sandstone surface.

strata. Visually distinct layers of sedimentary rock.

stratified. Layered.

stratigraphy. The study of physical and temporal relationships between rock units, as well as their environments of formation.

striation. One of multiple scratches or grooves, generally parallel, formed by the interaction of rock debris transported at the base of a glacier and the bedrock over which the glacier moves.

strike. The direction of a line formed by the intersection of an inclined planar geologic surface, such as a bedding plane, with the horizontal.

subduction. The process by which an oceanic plate descends beneath another tectonic plate.

symmetrical fold. A fold in which the axial plane is nearly vertical.

syncline. A downwarping fold.

talus. An accumulation of coarse, angular rock fragments at the base of a cliff.

tectonic. Pertaining to forces involved in the deformation of the earth's crust.

thrust fault. A reverse fault with a dip of 45 degrees or less.

till. Unsorted, unstratified sediment deposited by a glacier.

transgressive sequence. A sequence of rocks deposited during a rise in sea level.

tufa. A deposit of calcium carbonate that forms in brackish, mineral-rich water. Associated with pluvial lakes.

tuff. A general term for rocks formed of consolidated fragments of volcanic material, especially ash.

unconformity. A gap in the sedimentary sequence due to a period of erosion or nondeposition.

unsorted. Describes a sediment deposit that contains clasts of many different sizes.

uplift. Upward movement of the earth's crust.

upwarp. An arching of sediment layers.

uranium. Radioactive element that occurs in pitchblende, uraninite, and other minerals.

vent. The opening through which volcanic materials reach the earth's surface.

vesicle. A small cavity created by gas bubbles in igneous rock.

viscosity. A measure of a fluid's resistance to flow. The higher the viscosity, the slower the flow. Also **viscous**.

wave-cut terrace. A flat surface created by wave erosion.

weathering. The physical disintegration and chemical decomposition of rocks at the earth's surface.

welded tuff. A rock that forms from hot ash flows in which pieces of pumice join (or weld) together and flatten under the weight of overlying material.

xenolith. An inclusion of country rock within an igneous rock body.

SOURCES OF MORE INFORMATION

General Reading

Abbey, Edward. 1990. *Desert Solitaire*. New York: Simon and Schuster.

Chronic, Halka. 1990. *Roadside Geology of Utah*. Missoula, MT: Mountain Press Publishing Company.

Fillmore, Robert. 2000. *Geology of Parks, Monuments, and Wildlands of Southern Utah*. Salt Lake City: University of Utah Press.

Sprinkel, Douglas A., Thomas C. Chidsey, Jr., and Paul B. Anderson. 2000. *Geology of Utah's Parks and Monuments*. Utah Geological Association Publication 28.

Stokes, William Lee. 1986. *Geology of Utah*. Salt Lake City: Utah Museum of Natural History, University of Utah.

1. Snow Canyon State Park

Bugden, M. 1992. *Geology of Snow Canyon State Park, Washington County, Utah*. Utah Geological Survey, Public Information Series 13.

Sprinkel, Chidsey, and Anderson in "General Reading" above, 479–94.

2. St. George Dinosaur Discovery Site at Johnson Farm, and 31. Copper Ridge Dinosaur Tracks

Farlow, J. O., and M. K. Brett-Surman, eds. 1997. *The Complete Dinosaur*. Bloomington: Indiana University Press.

Thulborn, T. 1990. *Dinosaur Tracks*. London: Chapman and Hall.

3. Parowan Gap

Maldonado, F., and D. W. Moore. 1993. *Preliminary Geologic Map of the Parowan Quadrangle, Iron County, Utah*. U.S. Geological Survey Open-File Report 93-3.

Maldonado, F., and V. S. Williams. 1993. *Geologic Map of the Parowan Gap Quadrangle, Iron County, Utah*. U.S. Geological Survey Geologic Quadrangle Map GQ-1712.

Williams, Van S., and Florian Maldonado. 1995. Quaternary Geology and Tectonics of the Red Hills Area of the Basin and Range–Colorado Plateau Transition Zone, Iron County, Utah. In *Geologic Studies in the Basin and Range to Colorado Plateau Transition in Southeastern Nevada, Southwestern Utah, and Northwestern Arizona*, eds. R. B. Scott and W. C. Swadley, 257–275. U.S. Geological Survey Bulletin 2056-J:255–75.

4. Frisco and the Horn Silver Mine

Arrington, Leonard. 1963. Abundance from the Earth. *Utah Historical Quarterly* 31(3): 192–219.

Emmons, S. F. 1901. The Delamar and the Horn-Silver Mines: Two Types of Ore-Deposits in the Deserts of Nevada and Utah. New York. Vol. 31 of *Transactions of the American Institute of Mining Engineers.*

Hooker, W. A. 1879. *Report on the Horn Silver Mine.* Horn Silver Mining Company of Frisco, Utah, April 5.

Notarianni, Philip F. 1982. The Frisco Charcoal Kilns. *Utah Historical Quarterly* 50(1): 40–46.

www.utahhistorytogo.org/hornmine.html

5. Sevier Lake

Currey, D. R., G. Atwood, and D. R. Mabey. 1984. *Major Levels of the Great Salt Lake and Lake Bonneville.* Utah Geological Survey, Map 73.

Oviatt, C. G. 1989. *Quaternary Geology of Part of the Sevier Desert, Millard County, Utah.* Utah Geological Survey, SS-70.

6. The Virgin Anticline, Quail Creek State Park

Sprinkel, Chidsey, and Anderson in "General Reading" above, 465–78.

7. Hurricane Fault

Biek, R. F. 1998. *Interim Geologic Map of the Hurricane Quadrangle, Washington County, Utah.* Utah Geological Survey Open-File Report 361.

Stewart, M. E., W. T. Taylor, P. A. Pearthree, A. J. Solomon, and H. A. Hurlow. 1997. Neotectonics, Fault Segmentation, and Seismic Hazards along the Hurricane Fault in Utah and Arizona. *Brigham Young University Geology Studies* 42:235–54.

8. The Navajo Sandstone, Zion National Park

Sprinkel, Chidsey, and Anderson in "General Reading" above, 107–138, 607–618.

9. Weeping Rock, Zion National Park

Sprinkel, Chidsey, and Anderson in "General Reading" above, 107–138.

Welsh, S. L. 1990. *Wildflowers of Zion National Park.* Springdale, UT: Zion Natural History Association.

Williams, D. 2000. *A Naturalist's Guide to Canyon Country.* Helena, MT: Falcon Publishing.

10. The Springdale Landslide

Hamilton, W. L. 1992. *The Sculpturing of Zion.* Springdale, UT: Zion Natural History Association.

Hecker, S. 1993. *Quaternary Tectonics of Utah with Emphasis on Earthquake-Hazard Characterization.* Utah Geological Survey Bulletin 127.

11. Coral Pink Sand Dunes State Park

Sprinkel, Chidsey, and Anderson in "General Reading" above, 365–90.

12. Castle Rock Campground, and 13. Fremont Indian State Park

Francis, P. 1998. *Volcanoes: A Planetary Perspective.* Oxford: Oxford University Press.

Rowley, P. D., T. A. Steven, J. J. Anderson, and C. G. Cunningham. 1979. *Cenozoic Stratigraphic and Structural Framework of Southwestern Utah.* U.S. Geological Survey Professional Paper 1149.

Steven, T. A., P. D. Rowley, and C. G. Cunningham. 1984. Calderas of the Marysvale Volcanic Field, West Central Utah. *Journal of Geophysical Research* 89:8751–64.

Fremont Indian State Park and Museum Trail Guide. N.d. Salt Lake City: Utah Office of Museum Services.

14. Obsidian at Fremont Indian State Park

Janetski, J. C. 1998. *Archaeology of Clear Creek Canyon.* Provo, UT: Museum of Peoples and Cultures, Brigham Young University.

15. Fish Lake Valley

Hardy, C. T., and S. Muessig. 1952. Glaciation and Drainage Changes in the Fish Lake Plateau, Utah. *Bulletin of the Geological Society of America* 63:1109–16.

16. Sevier Fault

Reber, S., W. J. Taylor, M. Stewart, and I. M. Schiefelbein. 2001. Linkage and Reactivation along the Northern Hurricane and Sevier Faults, Southwestern Utah. In *The Geologic Transition, High Plateaus to Great Basin—A Symposium and Field Guide,* eds. M. E. Erskin, J. E. Faulds, J. M. Bartley, and P. D. Rowley, 379–400. Utah Geological Association Publication 30.

17. Bryce Canyon National Park

Sprinkel, Chidsey, and Anderson in "General Reading" above, 37–60.

18. Escalante State Park and the Petrified Forest

National Audubon Society. 1979. *Field Guide to Rocks and Minerals.* New York: Alfred A. Knopf.

Sprinkel, Chidsey, and Anderson in "General Reading" above, 411–20.

Wieland, G. R. 1935. *Cerro Cuadrado Petrified Forest.* Washington, DC: Carnegie Institute of Washington Publication 449.

19. The Waterpocket Fold, Capitol Reef National Park

Sprinkel, Chidsey, and Anderson in "General Reading" above, 85–106.

20. An Oyster Reef in the Desert

Peterson, F., R. T. Ryder, and B. T. Law. 1980. Stratigraphy, Sedimentology, and Regional Relationships of the Cretaceous System in the Henry Mountains Region, Utah. In *Henry Mountains Symposium,* ed. M. D. Picard, 151–70. Utah Geological Association Publication 8.

Robinson Roberts, L. N., and M. A. Kirschbaum. 1995. *Paleogeography of the Late Cretaceous of the Western Interior of Middle North America: Coal Distribution and Sediment Accumulation.* U.S. Geological Survey Professional Paper 1561.

21. Goblin Valley State Park

Sprinkel, Chidsey, and Anderson in "General Reading" above, 421–31.

22. Temple Mountain Uranium Mines

Dare, W. L. 1957. *Mining Methods and Costs, Calyx Nos. 3 and 8 Uranium Mines, Temple Mountain District, Emery County, Utah.* United States Department of the Interior, Bureau of Mines Circular 7811.

Hawley, C. C., D. G. Wyant, and D. B. Brooks. 1965. *Geology and Uranium Deposits of the Temple Mountain District, Emery County, Utah.* U.S. Geological Survey Bulletin 1192.

Kerr, Paul F., Marc W. Bodine, Dana R. Kelley, and W. Scott Keys. 1957. Collapse Features, Temple Mountain Uranium Area, Utah. *Bulletin of the Geological Society of America* 68:933–82.

Nininger, Robert. 1954. *The Uranium Ore Minerals.* New York: D. Van Nostrand Company.

23. The San Rafael Reef

Bauman, Joseph M., Jr. 1987. *Stone House Lands: The San Rafael Reef.* Salt Lake City: University of Utah Press.

Gartner, Anne E. 1986. *Geometry, Emplacement History, Petrography, and Chemistry of a Basaltic Intrusive Complex, San Rafael and Capitol Reef Areas, Utah.* U.S. Geological Survey Open-File Report 86-81.

Sprinkel, Chidsey, and Anderson in "General Reading" above, 421–31.

Tinkler, K. J., and E. E. Wohl, eds. 1998. *Rivers over Rock: Fluvial Processes in Bedrock Channels.* Washington, DC: American Geophysical Union.

24. Goosenecks of the San Juan River

Pederson, J. L., R. D. Mackley, and J. L. Eddleman. 2002. Colorado Plateau Uplift and Erosion Evaluated Using GIS. *GSA Today* 12(8):4–10.

Sprinkel, Chidsey, and Anderson in "General Reading" above, 433–48.

25. Valley of the Gods

Sprinkel, Chidsey, and Anderson in "General Reading" above, 529–34.

26. Natural Bridges National Monument

Sprinkel, Chidsey, and Anderson in "General Reading" above, 233–50.

Young, R. W., and A. Young. 1992. *Sandstone Landforms.* Berlin: Springer-Verlag.

27. Newspaper Rock

Castleton, Kenneth B. 1979. *Petroglyphs and Pictographs of Utah.* Vol. 2. Salt Lake City: Utah Museum of Natural History.

Cole, Sally J. 1990. *Legacy on Stone.* Boulder, CO: Johnson Books.

Jennings, Jesse D. 1973. *Prehistory of Utah and the Eastern Great Basin.* Salt Lake City: University of Utah Press.

Patterson, Alex. 1992. *Field Guide to Rock Art Symbols of the Greater Southwest.* Boulder, CO: Johnson Books.

Slifer, Dennis. 2000. *Guide to Rock Art of the Utah Region.* Salt Lake City: University of Utah Press.

28. Upheaval Dome

Alvarez, Walter, Erick Staley, Diane O'Connor, and Marjorie A. Chan. 1998. Synsedimentary Deformation in the Jurassic of Southwestern Utah—A Case of Impact Shaking? *Geology* 26:579–82.

Sprinkel, Chidsey, and Anderson in "General Reading" above, 619–28.

29. The Fiery Furnace

Sprinkel, Chidsey, and Anderson in "General Reading" above, 11–36.

Doelling, H. H., C. G. Oviatt, and P. W. Huntoon. 1988. *Salt Deformation in the Paradox Region.* Utah Geological and Mineral Survey Bulletin 122.

Nuccio, V. F., and S. M. Condon. 1996. *Burial and Thermal History of the Paradox Basin, Utah and Colorado, and the Petroleum Potential of the Middle Pennsylvanian Paradox Formation.* U.S. Geological Survey Bulletin 2000-O.

30. Arches National Park and the Entrada Sandstone

Baars, Donald L. 1994. *Canyonlands Country: Geology of Canyonlands and Arches National Parks.* Salt Lake City: University of Utah Press.

Sprinkel, Chidsey, and Anderson in "General Reading" above, 11–36.

Young, R. W., and A. Young. 1992. *Sandstone Landforms.* Berlin: Springer-Verlag.

31. Copper Ridge Dinosaur Tracks

See 2. St. George Dinosaur Discovery Site at Johnson Farm.

32. Round Mountain and the La Sal Laccoliths

Friedman, Jules D., and A. Curtis Huffman, eds. 1998. *Laccolith Complexes of Southeastern Utah: Time of Emplacement and Tectonic Setting—Workshop Proceedings.* U.S. Geological Survey Bulletin 2158:61–83, 85–100.

33. Periglacial Features in the La Sal Mountains

French, H. M. 1976. *The Periglacial Environment.* Essex, UK: Longman Scientific and Technical.

Giardino, J. R., J. F. Shroder, Jr., and J. D. Vitek, eds. 1987. *Rock Glaciers.* Boston: Allen and Unwin.

INDEX

But the love of wilderness is more than a hunger for what is always beyond reach; it is also an expression of loyalty to the earth, the earth which bore us and sustains us, the only home we shall ever know, the only paradise we ever need—if only we had the eyes to see. Original sin, the true original sin, is the blind destruction for the sake of greed of this natural paradise which lies all around us—if only we were worthy of it.

Edward Abbey, *Desert Solitaire*

ABOUT THE AUTHORS

Paleontologist and biologist **Bob Wieder** spent so much time in Death Valley, California, that friends christened him "Badwater Bob." Bob taught himself guitar on the tailgate of his pickup truck and now wanders the high plateaus of Utah, entertaining folks around the campfire and playing for tips in local bars.

Richard Orndorff, a faculty member in the Department of Geology at Eastern Washington University, is coauthor with Bob Wieder and Harry Filkorn of *Geology Underfoot in Central Nevada*, published by Mountain Press. He especially enjoys time spent in the outdoors with his wife, Karen, and daughter, Emma.

David Futey is a published author and photographer; he treks the wilds of Utah whenever possible, seeks out hikes at higher elevations for the good air, and enjoys deciphering the geologic stories behind landscapes. He spends his happiest moments with his wife, Susan, daughter, Mary, and son, Evan.

We encourage you to patronize your local bookstore. Most stores will order any title they do not stock. You may also order directly from Mountain Press, using the order form provided below or by calling our toll-free, 24-hour number and using your VISA, MasterCard, Discover, or American Express.

Some geology titles of interest:

_____Geology Underfoot in Southern California	14.00
_____Geology Underfoot in Death Valley and Owens Valley	18.00
_____Geology Underfoot in Illinois	18.00
_____Geology Underfoot in Central Nevada	16.00
_____Roadside Geology of ALASKA	18.00
_____Roadside Geology of ARIZONA	18.00
_____Roadside Geology of SO. BRITISH COLUMBIA	20.00
_____Roadside Geology of NO. AND CENTRAL CALIFORNIA	20.00
_____Roadside Geology of COLORADO, 2nd Edition	20.00
_____Roadside Geology of HAWAI'I	20.00
_____Roadside Geology of IDAHO	20.00
_____Roadside Geology of INDIANA	18.00
_____Roadside Geology of MAINE	18.00
_____Roadside Geology of MASSACHUSETTS	20.00
_____Roadside Geology of MONTANA	20.00
_____Roadside Geology of NEBRASKA	18.00
_____Roadside Geology of NEW MEXICO	18.00
_____Roadside Geology of NEW YORK	20.00
_____Roadside Geology of OREGON	16.00
_____Roadside Geology of PENNSYLVANIA	20.00
_____Roadside Geology of SOUTH DAKOTA	20.00
_____Roadside Geology of TEXAS	20.00
_____Roadside Geology of UTAH	20.00
_____Roadside Geology of VERMONT & NEW HAMPSHIRE	14.00
_____Roadside Geology of VIRGINIA	16.00
_____Roadside Geology of WASHINGTON	18.00
_____Roadside Geology of WISCONSIN	20.00
_____Roadside Geology of WYOMING	18.00
_____Roadside Geology of THE YELLOWSTONE COUNTRY	12.00
_____Finding Fault in California: _An Earthquake Tourist's Guide_	18.00
_____Fire Mountains of the West, Third Edition	20.00
_____Geology of the Lake Superior Region	22.00
_____Geology of the Lewis and Clark Trail in North Dakota	18.00
_____Geysers: _What They are and How They Work_	12.00
_____Glacial Lake Missoula and Its Humongous Floods	15.00
_____Living Mountains: _How and Why Volcanoes Erupt_	18.00
_____Northwest Exposures	24.00

Please include $3.00 per order to cover postage and handling.

Send the books marked above. I enclose $_____

Name_____

Address_____

City/State/Zip_____

☐ Payment enclosed (check or money order in U.S. funds)

Bill my: ☐ VISA   ☐ MasterCard ☐ Discover ☐ American Express

Card No._____Expiration Date:_____

Signature _____

MOUNTAIN PRESS PUBLISHING COMPANY
P.O. Box 2399 • Missoula, MT 59806 • Order Toll-Free 1-800-234-5308
E-mail: info@mtnpress.com • Web: www.mountain-press.com